Advances in Micro-Bioreactor Design for Organ Cell Studies

Advances in Micro-Bioreactor Design for Organ Cell Studies

Special Issue Editor

Carl-Fredrik Mandenius

MDPI • Basel • Beijing • Wuhan • Barcelona • Belgrade

MDPI

Special Issue Editor
Carl-Fredrik Mandenius
Linköping University
Sweden

Editorial Office
MDPI
St. Alban-Anlage 66
Basel, Switzerland

This is a reprint of articles from the Special Issue published online in the open access journal *Bioengineering* (ISSN 2306-5354) in 2018 (available at: http://www.mdpi.com/journal/bioengineering/special_issues/micro_bioreactor)

For citation purposes, cite each article independently as indicated on the article page online and as indicated below:

LastName, A.A.; LastName, B.B.; LastName, C.C. Article Title. *Journal Name* **Year**, *Article Number, Page Range.*

ISBN 978-3-03897-233-4 (Pbk)
ISBN 978-3-03897-232-7 (PDF)

Cover image courtesy of Danny van Noort.

Contents

About the Special Issue Editor

Carl-Fredrik Mandenius has been a professor in engineering biology at the Division of Biotechnology, Linköping University, Sweden, since 2000. He has previously worked in the pharmaceutical industry as director of PR&D and has been an associate professor at Lund University. His background is in biosensor technology and design and in bioprocess development, especially process measurement and control. Recently, he has been involved in several EU projects on stem cell development and toxicology, in particular towards the development and design of a new in vitro testing methodology using stem-cell-derived organ cells.

bioengineering

MDPI

Editorial

Advances in Micro-Bioreactor Design for Organ Cell Studies

Carl-Fredrik Mandenius

Division of Biotechnology, IFM, Linköping University, 581 83 Linköping, Sweden; carl-fredrik.mandenius@liu.se; Tel.: +46-13-28-1000

Received: 6 August 2018; Accepted: 9 August 2018; Published: 10 August 2018

The engineering design of microbioreactors (MBRs) and Organ-on-Chips (OoCs) has advanced considerably in recent years [1,2]. The term MBR originally referred to the bioengineering methodology of performing biological reactions in micro-scale reactor devices; the term OoC refers to the recreation of organs and tissues from the human body on or in a miniaturized device with a smaller volume than the original organ and with body-like fluids streaming around the cells in an in-vivo-like fashion. Despite this difference, many of the basic engineering principles coincide in the design of MBRs and OoCs. However, there are also striking differences in the design requirements due to the purposes of use.

In bioprocess development, the main aim of the use of MBRs is to accelerate the development work of new bioprocesses with microorganisms or mammalian cells as production organisms [3]. The culture of the manufacturing process is typically scaled down to 1–10 mL volume in the MBR, and critical process parameters and media composition are systematically optimized. The increased yield and productivity of the large-scale process can be reached at a much earlier stage in the development process with this approach. Commercial MBRs with >100 parallel MBR units are now on the market.

The aim of OoC devices is to facilitate the study of organ cell assemblies in vitro under conditions that recreate in vivo conditions of the organ in the body for recapitulating time-related cellular behavior [4,5]. An OoC device allows for the observation of cellular effects when exposed to drugs or other chemicals. This allows for the assessment of compounds' effects at subcellular and multicellular levels. Successfully applied, this supports the investigation of safety pharmacology.

This special issue addresses these diverse aspects of MBR design in nine expert contributions where a variety of cells and tissues are used with various aims and ambitions.

The fundamental design challenges in MBRs are highlighted in two review contributions [6,7]. The similarities of the design of different MBRs, despite purposes, lead to a general design methodology for MBRs where functionality drives the design [6]. The engineering-based design of MBRs and OoC devices can take advantage of established design science theory, in which a systematic evaluation of functional concepts and user requirements takes place. The review compares how such common conceptual design principles are applicable to MBR and OoC devices. The complexity of MBR design, which is exemplified for scaled-down cell cultures in bioprocess development and drug testing in OoCs for the heart and the eye, is discussed and compared with previous design solutions of MBRs and OoCs from the perspective of how similarities in understanding design from functionality and user purpose perspectives can more efficiently be exploited. The review can serve as a guideline and help the future design of MBR and OoC devices for cell culture studies.

Seldon and Fuller [7] further address the challenges of introducing organ and tissue cells in MBRs for understanding normal and pathological physiology. The differences and the constraints of the physiological environment that influence the design are highlighted. This review considers the key elements necessary to enable bioreactors to address the critical areas associated with biological systems.

The use of MBRs as tools for investigating tumor models is highlighted in two research studies [8,9]. Kuhlbach and colleagues [8] have studied tumor extravasation on a chip. Their device consists of three different parts, containing two microfluidic channels and a porous membrane sandwiched in between. In contrast to many other systems, this device does not need an additional coating to allow endothelial cell (EC) growth, as the primary ECs used produce their own basement membrane. The ECs in their device showed in vivo-like behavior under flow conditions. These results suggest that the new device can be used for research on molecular requirements and conditions and the mechanism of extravasation and its inhibition.

Toh and colleagues [9] have developed a microfluidic-based culture chip to simulate cancer cell migration and invasion across the basement membrane. In this microfluidic chip, a three-dimensional (3D) microenvironment is engineered to culture metastatic breast cancer cells in a three-tumor model. The chip is useful for drug screening due to its potential to monitor the behavior of cancer cell motility, and, therefore, metastasis, in the presence of anti-cancer drugs.

Investigating effects of drug compounds on organ cells in in vitro microfluidic models has been mentioned recurrently to fill the need in the pharma industry for more efficient drug testing. In a study by Christoffersson and colleagues [10], a 3D model with cells arranged in spheroids is shown to be a valuable tool to improve physiologically relevant drug screening. In this article, it is shown how the number of cells growing out from human-induced pluripotent stem cell (hiPSC)-derived cardiac spheroids can be quantified to serve as an indicator of a drug's effect on spheroids captured in a microfluidic device.

Freyer and colleagues [11] demonstrate another approach with the same purpose with liver cells. They investigated the response of primary human liver cells to toxic drug exposure in a miniaturized hollow-fiber-based bioreactor. The results validate the suitability of the microscale 3D liver construct to detect hepatotoxic effects of drugs in a perfused human in vitro culture platform.

In another MBR setup, Wrzesinski and Fey [12] carry out an in-depth study of hepatocytes metabolism. They describe basic principles and how they are regulated so that they can be taken into consideration when microbioreactors are designed. They provide evidence that one of these basic principles is hypoxia, a natural consequence of multicellular structures grown in microgravity cultures.

Aspects of fluid dynamics in MBRs are addressed by Tajsoleiman et al. [13]. Due to the sensitivity of mammalian cell cultures, understanding the influence of operating conditions during a tissue generation procedure is crucial. In this regard, a detailed study of scaffold-based cell culture under a perfusion flow is presented with the aid of mathematical modelling and computational fluid dynamics (CFD). The simulation setup provides the possibility of predicting cell culture behavior under various operating conditions and scaffold designs.

Another important aspect of MBR design is oxygen distribution. Fernandez et al. [14] demonstrate the use of oxygen sensors to measure the oxygen consumption rate of several variants during the conversion of styrene (substrate) to 1-phenylethanediol (product). The oxygen consumption rate allowed for distinguishing the endogenous respiration of the cell host from the oxygen consumed in the reaction. Furthermore, it was possible to identify the higher activity and different reaction rate of two variants relative to the wild-type NDO.

All together, these nine contributions reflect state-of-the-art aspects of MBR design and highlight the inherent potential and strengths of the concept of MBRs for organ cell studies.

Conflicts of Interest: The author declares no conflict of interest.

References

1. Bhatia, S.N.; Ingber, D.E. Microfluidic organs-on-chips. *Nat. Biotechnol.* **2014**, *32*, 760–772. [CrossRef] [PubMed]
2. Van Noort, D. Bioreactors on a chip. In *Bioreactors: Design, Operation and Novel Applications*; Mandenius, C.F., Ed.; Wiley-VCH: Weinheim, Germany, 2016; pp. 77–112, ISBN 978-3-527-33768-2.
3. Hemmerich, J.; Noack, S.; Wiechert, W.; Oldiges, M. Microbioreactor systems for accelerated bioprocess development. *Biotechnol. J.* **2018**, *13*, e1700141. [CrossRef] [PubMed]

4.	Khetani, S.R.; Bhatia, S.N. Microscale culture of human liver cells for drug development. *Nat. Biotechnol.* **2008**, *26*, 120–126. [CrossRef] [PubMed]

5.	Van Duinen, V.; Treetsch, J.; Joore, J.; Vulto, P.; Hankemeier, T. Microfluidic 3D cell culture: From tools to tissue models. *Curr. Opin. Biotechnol.* **2015**, *35*, 118–126. [CrossRef] [PubMed]

6.	Mandenius, C.F. Conceptual design of micro-bioreactors and organ-on-chips for studies of cell cultures. *Bioengineering* **2018**, *5*, 56. [CrossRef] [PubMed]

7.	Selden, C.; Fuller, B. Role of bioreactor technology in tissue engineering for clinical use and therapeutic target design. *Bioengineering* **2018**, *5*, 32. [CrossRef] [PubMed]

8.	Kühlbach, C.; da Luz, S.; Mueller, M.M.; Baganz, F.; Volker, C.; Hass, V.C. A microfluidic system for investigation of tumor cell extravasation. *Bioengineering* **2018**, *5*, 40. [CrossRef]

9.	Toh, Y.-C.; Raja, A.; Yu, H.; van Noort, D. A 3D microfluidic model to recapitulate cancer cell migration and invasion. *Bioengineering* **2018**, *5*, 29. [CrossRef] [PubMed]

10.	Christoffersson, J.; Florian Meier, F.; Kempf, H.; Schwanke, K.; Coffee, M.; Beilmann, M.; Zweigerdt, R.; Mandenius, C.F. A cardiac cell outgrowth assay for evaluating drug compounds using a cardiac spheroid-on-a-chip device. *Bioengineering* **2018**, *5*, 36. [CrossRef] [PubMed]

11.	Freyer, N.; Greuel, S.; Knöspel, F.; Gerstmann, F.; Storoch, L.; Damm, G.; Seehofer, D.; Foster Harris, J.; Iyer, R.; Schubert, F.; Zeilinger, K. Microscale 3D liver construct for hepatotoxicity testing in a perfused human in vitro culture platform. *Bioengineering* **2018**, *5*, 24. [CrossRef] [PubMed]

12.	Wrzesinski, K.; Fey, S.J. Metabolic reprogramming and the recovery of physiological functionality in 3D cultures in micro-bioreactors. *Bioengineering* **2018**, *5*, 22. [CrossRef] [PubMed]

13.	Tajsoleiman, T.; Abdekhodaie, M.J.; Gernaey, K.V.; Krühne, U. Efficient computational design of a cartilage cell regeneration. *Bioengineering* **2018**, *5*, 33. [CrossRef] [PubMed]

14.	Fernandez, A.C.; Halder, J.M.; Nestl, B.M.; Hauer, B.; Gernaey, K.V.; Krühne, U. Biocatalyst screening with a twist: Application of oxygen sensors integrated in microchannels for screening whole cell biocatalyst variants. *Bioengineering* **2018**, *5*, 30. [CrossRef] [PubMed]

bioengineering

MDPI

Review

Conceptual Design of Micro-Bioreactors and Organ-on-Chips for Studies of Cell Cultures

Carl-Fredrik Mandenius

Division of Biotechnology, IFM, Linköping University, 58183 Linköping, Sweden; cfm@ifm.liu.se;
Tel.: +46-13-28-8967

Received: 9 June 2018; Accepted: 14 July 2018; Published: 19 July 2018

Abstract: Engineering design of microbioreactors (MBRs) and organ-on-chip (OoC) devices can take advantage of established design science theory, in which systematic evaluation of functional concepts and user requirements are analyzed. This is commonly referred to as a conceptual design. This review article compares how common conceptual design principles are applicable to MBR and OoC devices. The complexity of this design, which is exemplified by MBRs for scaled-down cell cultures in bioprocess development and drug testing in OoCs for heart and eye, is discussed and compared with previous design solutions of MBRs and OoCs, from the perspective of how similarities in understanding design from functionality and user purpose perspectives can more efficiently be exploited. The review can serve as a guideline and help the future design of MBR and OoC devices for cell culture studies.

Keywords: biomechatronic design; bioprocess development; toxicity testing; in vitro assay; drug testing; heart-on-a-chip; eye-on-a-chip

1. Introduction

The engineering design of micro-bioreactors (MBRs) and Organ-on-Chips (OoCs) has attracted much attention in recent years [1–7]. The two terms, MBR and OoC, have diverse origins. The MBR derives its name from the bioengineering methodology of performing biological reactions in micro-scale reactor devices; OoC refers to the recreation of organs and tissues from the human body on or in a miniaturized device with a smaller volume than the original organ and with body-like fluids streaming around the cells in an in vivo-like fashion [8–10]. Despite this difference, many of the basic engineering principles coincide in the design of MBRs and OoCs. Probably due to that, the terms are used concurrently.

The kind of problems addressed with MBRs and OoCs are related. For bioprocess development, the MBRs are considered valuable tools for accelerating development of new bioprocesses with microorganisms or mammalian cells as production organisms [11,12]. The culture of the manufacturing process is scaled-down to 1–10 mL volume of the MBR, and critical process parameters and media composition are systematically optimized [13,14]. The MBRs for bioprocess development have even been scaled down to the size of chips (Bioreactor-on-a-Chip) [15] and used for mimicking chemostat or turbidostat bioreactors with bacteria and yeast [15–18]. The increased yield and productivity of the large-scale process can be reached at a much earlier stage in the process development with this approach. Commercial MBRs with >100 parallel MBR units are now on the market, e.g., ambr [19] and m3p systems [20].

With OoC devices, the aim is to facilitate the study of organ cell assemblies in vitro, under conditions that recreate in vivo conditions of the organ in the body for recapitulating time-related cellular behavior. The OoC device allows for the observation of cellular effects when exposed to drugs or other chemicals. This allows for the assessment of compounds' effects at subcellular and multicellular levels. Successfully applied, this supports the investigation of safety pharmacology and toxicology, and, when possible, the efficacy of drug compounds [7–10,21,22]. If OoCs generates reliable results, it will be important to control the cells' survival in the artificial milieu of small-scale OoC; their stability relates to the design of liver devices [23–25]. However, other organs [26–29] have attracted almost the same interest, e.g., pancreas, eye, cartilage, heart, and lung cells [28,29], as well as combinations of several organ cell types on the same chip [30,31]. This is of great interest, as are tumors-on-chips, which are also covered in this special issue [32,33].

Importantly, the possibility to correctly observe, monitor, and analyze the effects of the cells through imaging, sensors, and other analytical means, as well as by controlling the process in the process development MBR or OoC, are pivotal for all applications of MBRs. Miniaturized sensors, optical fibers, and imaging inside or at the outlets of the MBRs/OoCs may provide these opportunities in a reliable way [34–37]. The design of the fluidics and transport inside the reactor chamber and the mixing in the device are common design problems. This results in some very similar design solutions.

Figure 1 illuminates the similarities and diversities of MBR and OoC device for various applications. Although the intended use and outcome of the devices differ (Figure 1A), the transformations by the biological components in the devices are similar (Figure 1B). The common transformation process that occurs in every MBR device provides similar prerequisites for the design. The resulting design solutions, such as a rack of small MBR-containers with optical sensors placed at the bottom of each micro-vessel (Figure 1C); the compact artificial liver bioreactors with intertwined hollow-fibers for liquid and gas transport (Figure 1D); or small channels with an internal membrane for transepithelial electrical resistance measurement (TEER) for drug penetration studies, PDMS chips with double channels, and parallel channels (Figure 1E) are all examples that share the general structure in Figure 1B. Consequently, the design of the devices should follow that frame.

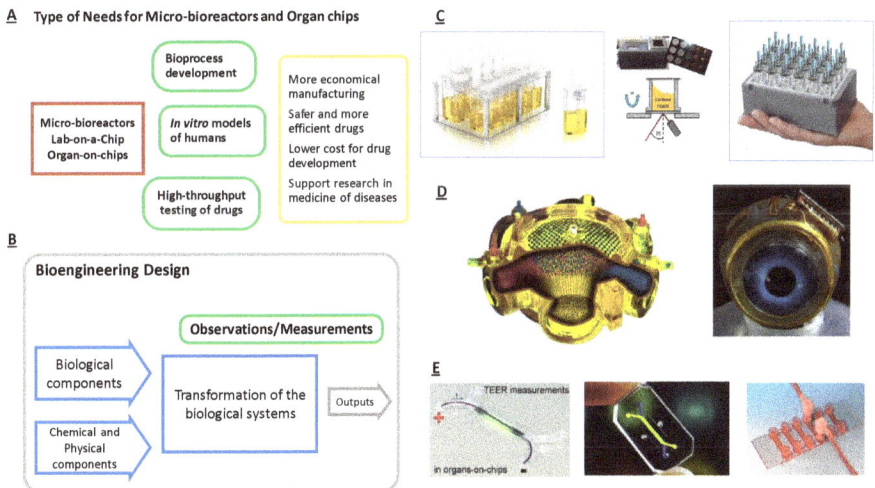

Figure 1. The similarities and diversities of micro-bioreactor design. (**A**) MBRs and OoCs have distinct purposes, which are expressed as user needs and requirements, e.g., the device should generate data for more economical manufacturing, safer and more efficient drugs, low cost for development, and support for medical research; (**B**) Despite the diversity in needs, a common transformation process can be outlined; (**C**) This is shared for MBRs used for process development; (**D**) for artificial liver bioreactors; and (**E**) for small plastic chips with multi-channels and internal membranes.

Thus, the engineering design of MBRs and OoCs have much in common, especially at a conceptual level. Established conceptual design methodology [38,39] could therefore significantly facilitate the development process of new MBR and OoC devices. The established conceptual design methodology is based on approaching the design of a new product from a functional perspective in which the functionality of the product drives the development of the design. In industrial design, conceptual methodology is widely applied to mechanical and electrical products [40]. However, in bioengineering it has been so far rarely used, with only a few examples on bioreactor scale-up [41,42], bioprocess configuration, monitoring and control [43,44], and stem cell production [45,46], but also recently for organ-on-chips [47].

In this review article, it is shown how the general conceptual design methodology can be applied to develop and improve the design of micro-bioreactors on a functional level. Two examples of conceptual design are shown to illuminate the similarities and differences in the design of MBR and OoC: (1) an MBR for production of hamster cells and (2) a heart-on-a-chip reactor for drug testing.

2. Conceptual Design Methodology

The conceptual design methodology is based on a systematic procedure to analyze the design objectives and, from these, conceive alternative design solutions that meet the objectives for a new product prototype [38–40,48]. The workflow in the development process in conceptual design starts with identifying the functions that are required to realize the user needs of the product and, from that, select and configure functional components that can effectuate user needs (Figure 2). Alternative configurations are compared with user needs and ranked versus user needs. This results in a preferred configuration. Once this choice is made, the functional components of the configuration are replaced with real physical components or objects. This results in a blueprint for an initial prototyping, which then undergoes testing and is transferred to the manufacturing of the product [30]. The methodology is well-known in mechanical engineering. It is seldom applied in bioengineering, e.g., for bioprocesses, biosensors, and organ chip device design. In the following, the general workflow in conceptual design when developing a MBR prototype is described.

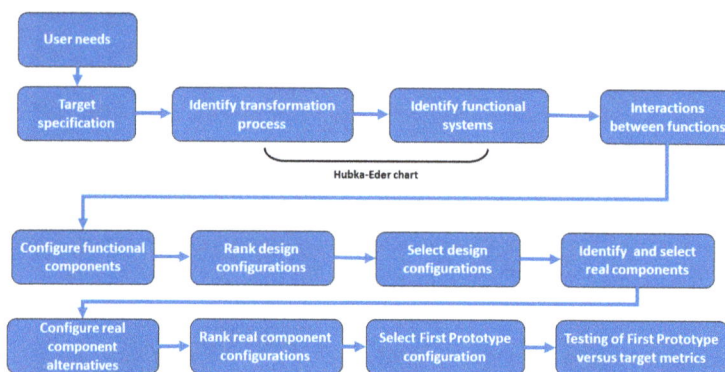

Figure 2. The workflow in conceptual design when developing a new product prototype as suggested by several authors [30–32]. The steps are iterative and partly parallel to speed up the work in the design team.

2.1. Design Objectives, User Needs and Specification of Target Values for Design

For achieving success with the conceptual design, a stringent and covering description of the design objective, or the design mission, is the starting point in the process of finding the appropriate design solution. The description of the objective constrains the design. Indirectly, it also encircles existing or potential users [49]. Once the users are known, they can be interrogated about their actual needs and requirements on the design solution [50]. Generally, the user needs for MBRs and OoCs vary with the purpose of the targeted product, which results in different priorities as shown in Table 1. The table elucidates the wide variation in character of needs and requirements that exist, in which some significantly impact the design, while others do not. For example, a certain number of cells are necessary for recapitulating an in vivo function, which sets the minimal size of the MBR-unit; critical biomarker molecules (or protein product, side-products) are produced in amounts possible to measure, which sets another minimal limit for the number of cells required in the MBR-unit; distribution of oxygen and uptake of oxygen in the unit may also be requirements that constrain the design.

Table 1. Examples of common user needs of micro-bioreactors and organ-on-chips.

Users' Needs and Requirements	Rationales or Examples	Priority
Maintain human organ cells with number of a corresponding human organ equivalents	Lower number of cells in a reactor unit would not show relevant data	10
Cells shall maintain the same functionality in vitro as they do in vivo	In most assays, the functionality is the target end-point to be observed	10
A multi-cellular system should be recapitulated in the MBR/OoC	A human organ or tissue is in vivo, like if it interacts with adjacent cells	9
Excreted metabolites shall be analysed in situ or at line with sensitive analytical means	Amounts of analytes produced in the MBR are minutes due to scale	9
Cell densities equivalent to an industrial production system	In-process development should have cell concentrations on a large scale	8
Material properties of MBR should not interfere with the biological transformation	Some materials are toxic, absorb drugs, or affect gradients of gases	8
Microfluidic conditions in the device should not harm the cells capacity	Shear force in the micro-reactor shall not change cells functional behaviour	7
The MBR/OoC shall be operable with stable performance over extended time periods	Short-term acute effects (<1 day) are of lesser value than chronic (>14 days)	7
Compounds should be exposed to cells or cell organoids in a relevant way	Diffusion, shear, and gradients in MBR should reproduce in vivo perfusion	6
Allow controlled addition of media factors	The exposure of factors	2
Gradients in the MBR of O_2, CO_2, pressure, and temperature should be in vivo-like	Variations in gradients are known to influence cellular response	1

Other more specified user needs for an MBR related to the design objective could, for example, involve longevity of use of the MBR, generation of gradients in pO_2 and nutrients in the device, transformation rates of the cell culture, scalability of units, and flow-through rates. Such needs are highlighted by other authors in this special issue, e.g., by Wrzesinski & Fey [24] on oxygen supply to liver cells in an MBR, by Freyer et al. [23] on 3D microstructure of liver cell MBRs, and by Fernandez et al. [37] on oxygen measurement in MBRs.

Once needs are identified and clearly described, they are also specified with target values or range of values. The target metrics may vary considerably. Table 2 gives examples for a liver-on-a-chip [47]. Note that specified values can be either quantitative or qualitative.

Table 2. User needs of a micro-bioreactor with organ cells.

User Needs	Target Metrics (Examples)	Specification (Examples)
MBR shall have multi-cellular systems	At least three cell types	Hepatocytes, Kupfer cells, fibroblast
Cells shall have in vivo-like functionality	Cells per MBR unit	50,000–75,000 cells
Cells as in human organ equivalent	Cells in equivalent	25,000 cells
Extracellular matrix	Hydrogel type	Matrigel or RGD-PEG
Flow of nutrients	Shear force number	
Measurement be extended time periods	Days	>10 days
Controlled addition of growth factors	Pump rates	
Microscope In situ inspection	Microscopic resolution	±10 nm
Sampling of effluent fluid	Number sampling ports	3
Oxygen transfer	Dissolved oxygen tension	More than 10%
Oxygen permeability of device material	mg O_2 per mL and hour	
Material properties of device	Porosity	25–30%
Recycling of media	Recycling ratio	1–2

2.2. Mapping of the Functional Systems of the Transformation Process and Assessing Their Interactions

A key activity in conceptual design is to describe and establish the structure of the transformation that the designed prototype should perform and what functions are required to perform the transformation. This is done in a graphical representation of the transformation and functions (Figure 3A), a so-called Hubka-Eder map (named after Vladimir Hubka (1924–2006) and Wolfgang Ernst Eder (1930–2017), the originators of this representation) [38]. The essential purpose of this map is (1) to define the transformation process, usually in the phases for preparation, main transformation, and finishing, that should take place in the designed device; and (2) to define the functions required for carrying out the transformation process [48]. The functions are structured into groups of systems, and these are further broken down into functional subsystems (Figure 3B). In a mathematical formalism, this can be described as

$$R = \sum_{i=0}^{n} \times \left(\sum_{j=0}^{n} FSi, j \right)$$

in which FSi,j are the functional sub-systems. The i index refers to main functional systems, which we divide into the biological systems, the technical systems, the information systems, the management systems, and the human systems necessary for carrying out the transformation process. To these systems, we also add the unknown surrounding environment, referred to as the active environment, which can influence the functions in ways we are not able to foresee. That could be biological variation, unpredictable sample background, or even influences associated with laws and regulations.

The functional systems and sub-systems interact with the transformation process to drive it forward but can also interact with adjacent functional sub-systems. Understanding the effects of these interactions on the transformation process is the core for accomplishing a functional prototype and is fundamental for making important design decisions. Figure 3C shows a convenient way to represent these interaction effects by ordering them into an interaction matrix, IM:

$$IM = \begin{bmatrix} w1,1 \ FS1,1 & \cdots & w1,j \ FS1,j \\ \vdots & \ddots & \vdots \\ wi,1 \ FSi,1 & \cdots & wi,j \ FSi,j \end{bmatrix}$$

in which the functional sub-systems are represented by a position in the matrix. The weights wi,j are values estimating the interaction strengths. These strengths are not precise measures and should only be seen as relative estimates for comparing each FSi,j.

When designing MBR devices, technical data from literature can often be helpful. The wealth of published data should, however, be carefully used and valued.

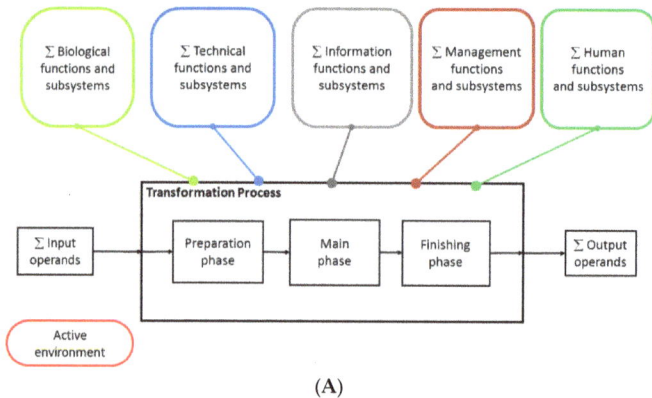

(A)

Figure 3. *Cont.*

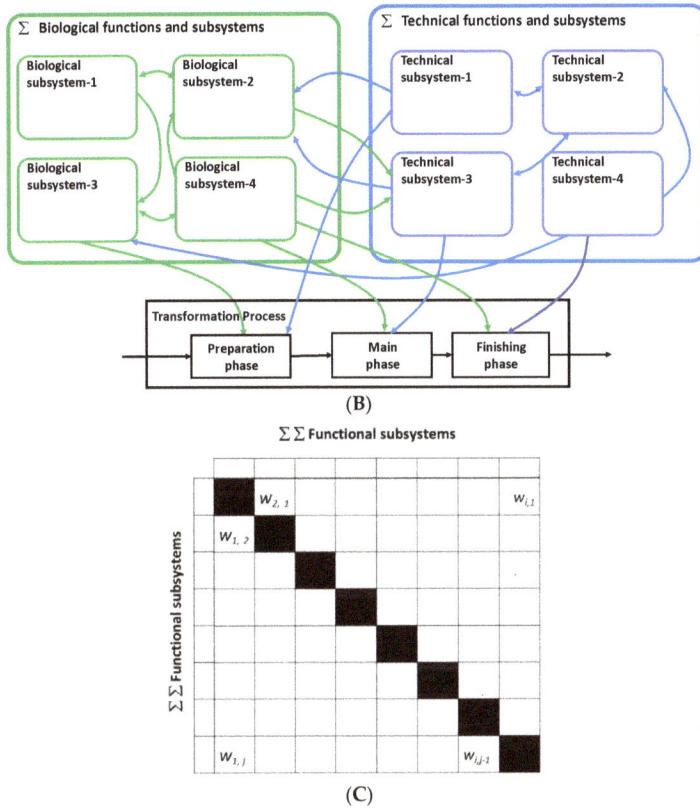

Figure 3. (**A**) Overall Hubka-Eder map showing the transformation process and functional systems; (**B**) a zoom-in of the biological and technical subsystems and their interactions in between and with the transformation process phases; and (**C**) the interaction matrix with assessed interaction effects between subsystems.

2.3. Key Functional Components

The Hubka-Eder mapping and the interaction analysis facilitate identification of essential functional components necessary for the design of the prototype [38]. Figure 4A shows a collection of 17 functional components suitable for the design of an MBR prototype, with the functional components grouped in the function systems from the HE-map. Note that a functional component is solely a conceptual object capable of carrying out the function, not a defined physical component, such as a valve or a pump. For example, a fluidic transporter only tells what you want the component to do, not if it is produced by pumping, using a syringe, or utilizing gravitational force. These "neutral", non-physical, or non-real components allow one to investigate how the functions can be configured. Figure 4B shows two examples of configurations from the 17 functional components. In one configuration (I), cells are transferred to a temperature-controlled space in which the biological systems are preserved in several separately contained and temperature-controlled units with sensor functions. In the other configuration (II), all units are placed in the same temperature-controlled unit, and sensors are shared between the contained cell units. Theoretically, the 17 functional components in Figure 4 can be combined in a multitude of configurations, most of them unrealistic, but a few are realistic and worth investigating further.

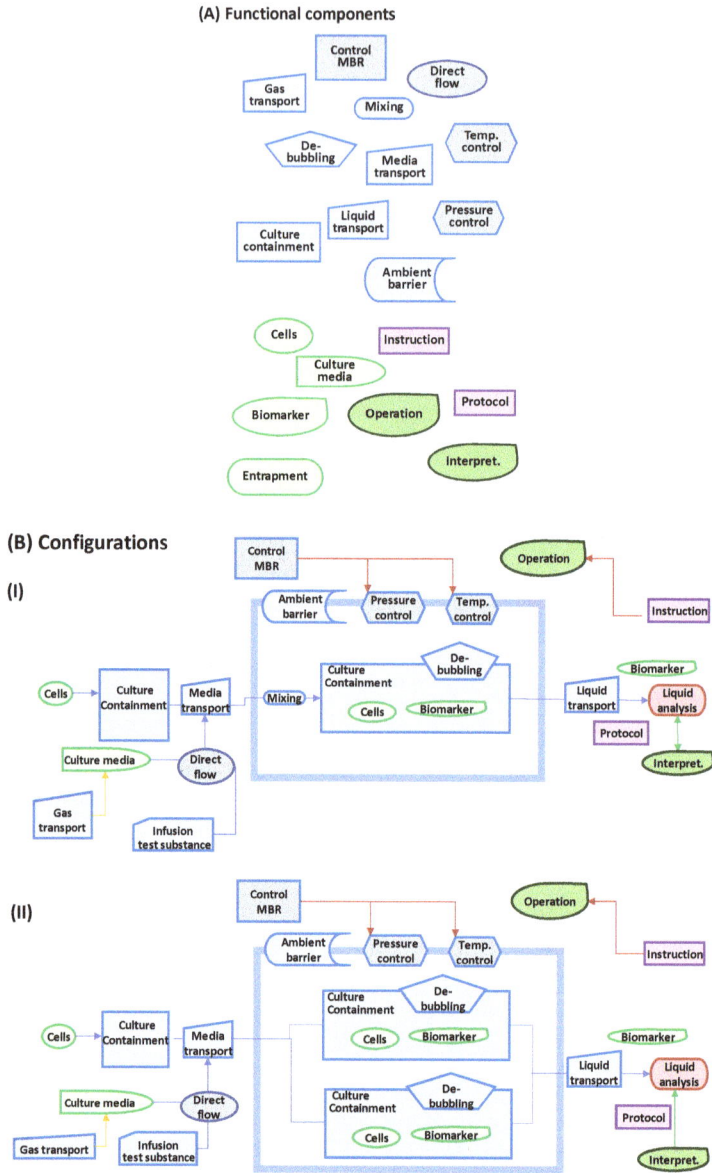

Figure 4. (**A**) Biological (green), technical (blue), and information (red) functional components required in an MBR device as identified from the functional subsystems in the HE-map in Figure 3; (**B**) Two examples of configurations from these functional components in which two MBR units are included in the prototype design.

2.4. Assessment of Configuration Alternatives

The configuration alternatives generated from the functional components should now be compared and assessed versus the user specification target values in Table 2 [40]. Initially, a relatively high number of configuration alternatives can be assessed by a rough screening (20–30 configurations depending on number of components). By that, the number of configurations can be reduced to less than 10, and these can be assessed more thoroughly. Typically, the configuration alternatives can be limited to five or six. We rank their assumed effect on the target specification value; either this is quantitative or qualitative. The ranking is at this stage relative, but could be quantified more exactly, e.g., variability, standard deviation, limit of detection, or analysis time. This will require experimental evidence, e.g., measurements in a test bench in a pre-prototype. Table 3 shows an example of estimated ranking scores for four configuration alternatives. A four-level ranking as shown in the example is sufficient to discriminate the configurations versus the specification values. In Table 3, alternative 2 gets the highest score and is chosen for further development towards a prototype. In the table it is also possible to introduce weight factors to tune the importance of each need in relation to the design objective. For example, low price may have much more impact than size. The balance of the weights is decisive for the final ranking score. A design team must consequently be aware of this and be cautious of how they treat values. However, if they do this, it will become an efficient means with which to perform sensitivity and risk analysis at this early stage of the design.

Once the functionally most feasible design alternative has been selected from the scoring of the user needs, the real physical components replace the functional components [48]. These real components are identified and chosen in a similar selection process. The real component alternatives are screened and scored versus the user needs. Thus, the columns in Table 3 represent real component alternatives that are compared versus the listed user needs and scored.

The real components can be collected from commercial vendors or be specially designed by the team. Of course, commercial components are preferred when they are available and specifications are known. The screening is done similarly and results in a preferred alternative. This alternative will then be prototyped and tested.

Table 3. Example of ranking of design alternatives versus target specification metrics [1].

Design Alternatives User Needs	Alternative 1	Alternative 2	Alternative 3	Alternative 4
Allow co-culture of cells	••	•••	••	•
Cells have in vivo-like functionality	••	•••	•	•
Number of cells in device as in an in vivo equivalent	•	••	••	•••
Extracellular matrix possible to mimic	•••	•••	•••	•••
Continuous flow of nutrients	••	•••	••	•••
Measurement periods up to 3 weeks	-	•	•••	-
Exposure of test compounds to cells	•••	•••	•••	•••
Allow controlled addition of growth factors	•••	•••	-	••
In situ inspection with confocal microscope without interference	•••	•••	-	-
Sampling of effluent fluid	•••	•••	•••	•••
Oxygen transfer through device	•	•	•••	•••
Liquid permeability of device	••	••	••	-
Device shall allow recycling of media or exposed compounds	•••	•••	•••	•••
Material properties of device not interfering	•	•	•	••
Recycling of outlet flow	•••	•••	-	•
Total score of ranking	32	37	28	28

[1] Compliance with user needs ••• high; •• partly possible; • low or uncertain; and - none or impossible.

3. A Micro-Bioreactor for Process Development of Monoclonal Antibody Production

The design mission in this application is to develop an MBR prototype that can be used as a tool in process development of large-scale bio-production of monoclonal antibodies [17,51–54]. One of the dominating products on today's biotechnology market is recombinant monoclonal antibodies for use as bio-therapeutics and for diagnostics. The most commonly used production organism is Chinese Hamster Ovary (CHO) cells, but also other cells are used, such as hybridoma and HEK cells [55,56]. Before scaling-up to production, the culture conditions are thoroughly investigated in process development on a small scale [57]. This ensures that the production process is optimized with regards to cell growth rate, antibody production rate, and culture media composition; has optimal values for physiochemical parameters such as temperature, pH, and dissolved oxygen tension in the reactor; is optimal with regard to initiation and propagation of the recombinant expression system in the CHO-cell culture; and applies feeding profiles of nutrients and other growth factors to the bioreactor [58]. In the case of monoclonal antibody production, great concern is devoted to reducing formation of variants of the IgG molecule, such as multimeric and fragmented forms or various glycosylated forms of IgGs [59]. For other bioprocesses, with other expressed proteins, similar modifications are of concern, such as oxygenated, aminated, clipped, or degraded product molecules. All these process development issues are very time-consuming tasks, but are absolutely necessary to scrutinize in the early process R&D.

The MBRs are excellent tools for such R&D work due to their small size and possibility for parallel testing. By using MBRs process, development can be accelerated significantly [11]. Examples of successful steps taken in this direction are emerging mini- or microscale bioreactors with online sensors for measurement of critical process parameters [19,20]. These scaled-downed bioreactor systems have volumes in the range 1–20 mL. Commercial systems are already on the market for suspension cell cultures with a variety of designs [24] (see also Figure 1C).

Conceptual design methodology can efficiently support the design of new MBRs for these purposes [43]. Bioprocess developers have good practical notions of the needs and requirements of MBRs for process development purposes. The most common requirements and specifications for efficient scale-downed optimization work are shown in Table 4.

Table 4. User needs of a micro-bioreactor for process development of mammalian cell cultures.

User Needs	Target Metrics	Specification
Biological functions		
Mammalian cells shall be used	Cell type	CHO, HEK cells
Concentration range of cell culture	Cell/mL	10,000–10,000,000
Expression of extracellular protein	Proteins expressed	IgGs
Same culture media shall promote both growth and expression	Type of media to be used	Serum-free medium
Culture time	Days	7–14 days
Technical functions		
Gentle well-distributed mixing	Shaken or stirred	Shaken
In situ inspection with confocal microscope without interference	Yes/No	Yes
Sampling of effluent fluid	Offline/inline	Offline
Oxygen transfer	$k_L a$ value for OTR	$>100\ h^{-1}$
Permeability of device	Oxygen permeability (%)	<1%
Material properties of device	Surface hydrophobicity (angle)	10 degree
Information functions		
Online information about physical conditions in the MBR	Sensor types	Temp., pH, pO$_2$
Offline information about content of culture media	Analytes analyzed offline	All monomers in culture media
Offline information about IgG forms	Analytes analyzed offline	IgG forms
Low fabrication cost	Percentage of the sales price	>10%

The table addresses the wide range of cell density or cell number required. It is critical in an MBR to have a volume that is large enough for generating statistically trustworthy data, achieving homogenous fluid in the miniaturized bioreactor system and creating gradients comparable to the scaled-up system. This demand complicates the design for transfer of gaseous molecules, in particular, for oxygen transfer from the gas bubble phase to the liquid phase at the scale of the MBR. Previous designs have either been designed as microtiter-like MBR arrays with sensors in the bottom of each bioreactor-well (e.g., Sartorius, m2p-labs) [19,20] or as separate MBR units with impellers and submerged micro-sensor probes and external pumps (Ambr) [19]. Comparisons with laboratory scale bioreactors of 2–5 L show good correlations, thus indicting that homogeneity of the liquid in the MBR and laboratory scale reactors at this scale is the same when testing mammalian cells, which are the easier cases due their low transfer rate [60]. The table addresses conditions related to these rheological functions of the MBR, as efficiency of mixing of the fluid, transfer of oxygen, addition and withdrawal of nutrients, and sterility demands. These requirements coincide with large-scale cell culture reactors and have also previously been discussed [43].

Furthermore, the cell growth and protein expression in the cell culture in the MBR need to be monitored to provide information about the culture for optimizing conditions and procedures to be transferred to the large-scale process [61,62]. This includes online monitoring of temperature, pH, and dissolved oxygen and offline measurements of components in the bioreactor media, such as product forms, excreted metabolites, and residual nutrients. Finally, the requirements on the MBR design should address management protocols for data analysis and operation, as well as specifications for MBR fabrication (materials, cost, and maintenance needs).

The specification of the user requirements in Table 4 suggests a Hubka-Eder map with function systems as depicted in Figure 5. The biological system functions follow from the design mission where the cell line (CHO-cells) has the prime function to express a high titer of the target product (monoclonal IgG) using a recombinant gene construct. The choice of a suitable CHO cell line and expression system is decisive for reaching the production goal. The composition of the culture medium is decisive for the performance of the culture in respect to growth rate, as well as product formation rate. Thus, the interactions between these subsystems, the cell line, the expression system, and the medium are critical issues for the design of the MBR.

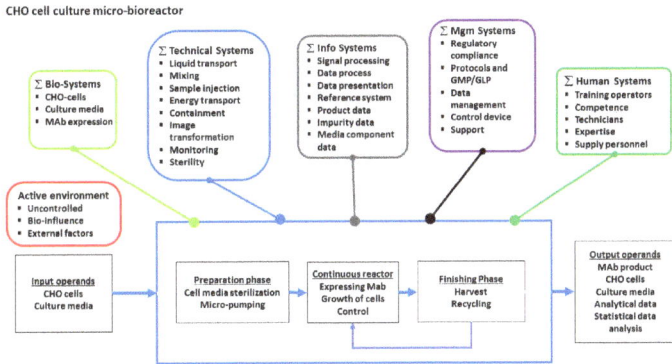

Figure 5. Hubka-Eder map showing the functional systems and subsystems in a continuous recycled MBR for bioprocess optimization of a CHO cell culture producing monoclonal antibody.

The technical system functions are the following: (1) to contain the miniaturized cell culture volume; (2) to provide a sterility barrier towards the environment; (3) to aerate the culture with oxygen a way that is comparable with the large-scale application; (4) to transport culture media with nutrients and excreted product(s) to and from the contained culture; and (5) to measure in real-time the physical and chemical conditions of the culture and its media [57].

The functions of the information system are to collect information from real-time measurements and offline analysis performed in the effluent from the MBR and display these data for process optimization. The information system should also provide information about the performance of the MBR itself and its control functions for temperature, pH and pO_2, and transport systems (e.g., mixing and pumping).

The functions of the management systems are to provide protocols for MBR operation, computation methods for data analysis, and control of MBR operations through its technical systems. The human systems involved are the laboratory technicians that operates the MBR, the process development engineers and other experts that compute and interpret data from MBR experiments, and service and support personnel that maintain the MBR. These humans interact with all of the other systems and must be able to do that efficiently.

Also shown in the HE-map in Figure 5 is the active environment. This could, for example, be unanticipated biological variations of the performance of the cell line and expression system, partly ascribed to our lack of biological understanding of the complexity of the cell machinery of the CHO cells. Such influences or other influences from the surrounding environment of the MBR should be compensated for, especially by the technical subsystems based on data from the information systems.

All of the systems in the HE-map interact with each other and with the phases of the transformation process. The strengths of these interactions should be analyzed in the interaction matrix (cf., Figure 3C), and those interactions that have high impact should be singled out.

Based on this analysis, 16 functional components can be identified that need to be available when configuring the design. Figure 6A shows the functional components for containment, mixing, infusion of media, real-time data sampling, and transportation. The components of the biological functions are limited to the function of the cell line to express IgG, the gene function components, and the functions of the culture medium to enhance growth and IgG production. Information components are the information *per se* and the devices that can generate the information. The management components are software and instructions and rules to be followed. The humans are the individuals able to carry out the functions. These 16 components can easily be arranged in more than ten plausible and realistic configurations. Figure 6B shows two of these configurations, both close to existing MBR products [19,20].

In a full design analysis at least up to 20 configurations would be arranged from the functional components. In configuration I in Figure 6B, each of the parallel MBR units is controlled separately with its own sensors and control units, although they are supervised with one management system. In configuration II, the MBR units are placed in one sensor and control unit. Importantly, the configuration diagram shows only functions for which several physical objects or devices may effectuate the functions.

Comparing these two configurations with the specification list in Table 4 results in a ranking that does not favor any of the configurations (Figure 6C). In the ranking, the weight factors for each specified need have been the same ($w = 1.0$). This may, however, not reflect the actual importance of each need. For example, if the cost is upgraded to $w = 2$, which is a sound assumption, and the other need attributes are tuned down, the total score ranks makes the second configuration (II) alternative the preferred design solution.

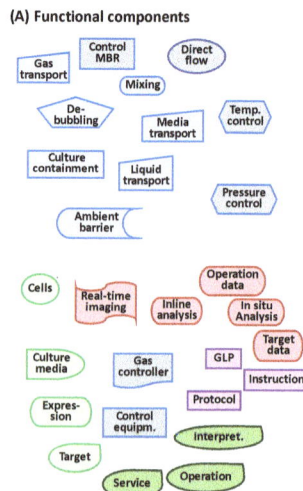

Figure 6. *Cont.*

(B) Configurations

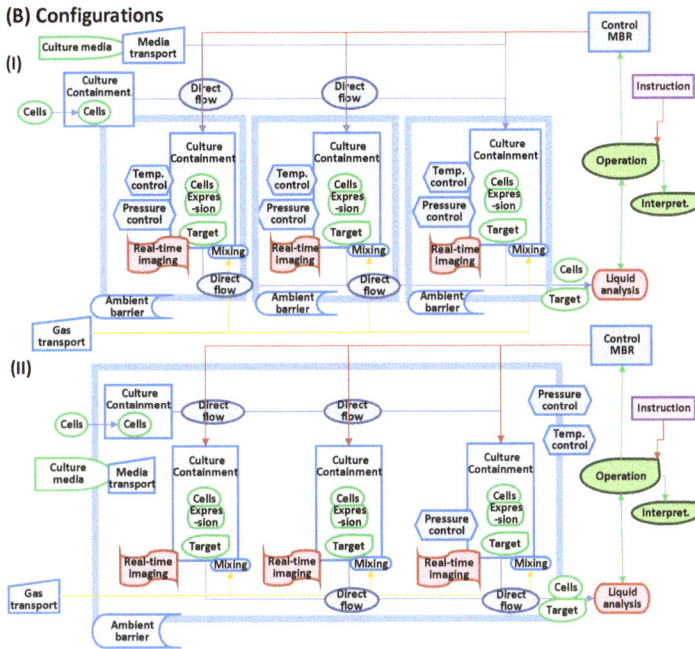

(C) Scoring of user needs (MBR in process development)

	User needs	Configuration I	Configuration II	Weights
1	MBR allows cell density of 0.1 10⁶ – 100 10⁶ cells/mL	4	4	50%
2	MBR shall remain sterile for at least 30 days experimentation	4	5	40%
3	Chamber of MBR shall not have any gradients	4	4	80%
4	Mixing efficiency of MBR should reproduce large scale	4	4	90%
5	MBR shall allow continuous and fed-batch operation	5	3	80%
6	MBR should have as many real-time sensors as possible	3	4	70%
7	MBR shall be easy to scale-out for high-throughput tests	3	5	60%
8	MBR shall be convenient to operate for lab technicians	4	4	40%
9	Fabrication cost of MBR shall be as low as possible	2	5	200%
	Score with 100% weights	38	38	
	With weight factor as in column	*23.9*	*30.6*	

Figure 6. (**A**) Functional components for conceptual design of a MBR for bioprocess development; (**B**) Two configurations of these functional components; (**C**) Comparison and ranking of these two configurations versus user needs in Table 4 in which scores are calculated for weight factors all 100% or as in last column in the table.

In the next step of the design, the chosen conceptual design alternative is translated into a configuration with real physical components replacing the functional components. Thus, common components, such as plastic containers, micro-pumps, temperature and pH sensors, impellers, or robotic racks are introduced and compared. In a second ranking table, the need specifications are used again and component alternatives scored (pumps, gravity mixers, osmotic pumps, and optical and electrical sensor devices) according to their ability to meet the requirements of the specifications (see ranking Table S1 in Supplementary Information) [63–67]. With these ranked components, a first prototype is blueprinted, constructed, and tested (Figure 7). This prototype undergoes testing procedures as commonly done in all product development in the manufacturing industry (cf. detailed description in e.g., Ulrich & Eppinger [40]).

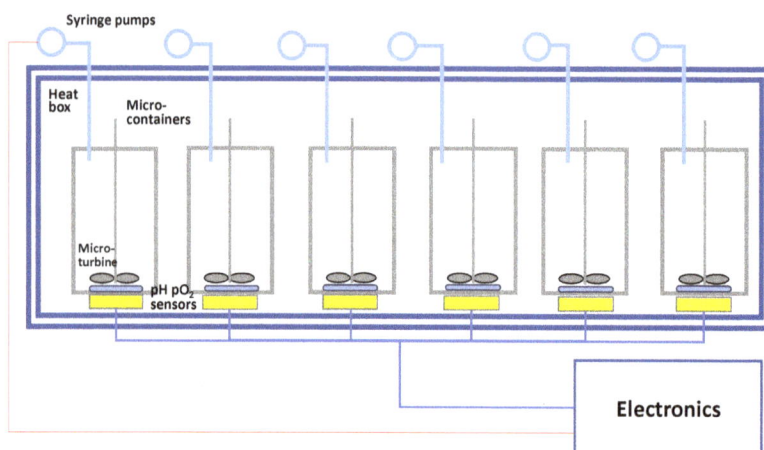

Figure 7. Blueprint of a prototype based on the ranking of components towards the need specifications. An MBR setup with six independent reactors for culture of CHO cells in 2–5 mL scale contained in a thermostat box. Sensors for pH and pO$_2$ measure online in each MBR.

4. A Heart-on-a-Chip Micro-Bioreactor for Assessment of Drug Safety and Efficacy

Several designs of heart-on-a-chip (HoC) devices for assessment of clinical in vitro heart models [68–70] and for testing of toxicity and efficacy effects of drug substances on cardiac cells have been presented [71–77]. These studies have shown that the HoCs devices have the potential to facilitate and shorten drug development in the pharmaceutical industry when the HoCs can accurately reflect relevant in vivo conditions in human heart tissues related to critical drug effects. The challenge is to accomplish theses in vivo-line conditions in the HoCs. In this respect, the access to induced pluripotent stem cells (iPSC) from stratified patient groups with different genetic background makes iPSCs an attractive approach that bypasses animal testing with lesser relevance [78]. These prerequisites and ambitions have guided most recent efforts in designing HoC devices. Examples are studies with electrical or visual recording of the heart's beating rate with cardiac bodies [78] exposed to drug substances at varying concentrations [79].

In the example shown here, the design objective is to design a HoC prototype using commercially available, iPSC-derived cardiomyocytes [80] or cardiomyocytes developed in their own lab, e.g., at a research institute or a R&D unit at a drug company [81]. The purpose of the design of the HoC device is to assess effects of drug candidates, either for cardiac chemotherapy or other therapies in which the heart is affected by the drug.

The cardiomyocytes in the HoC prototype will experience the same spatiotemporal conditions as in the in vivo heart tissues. This requires a 3-dimensional cellular microstructure inside the HoC flow-through chamber, in which other cells in the cardiac tissue shall be included as well. This could either be assemblies of co-cultured cells or assembled scaffold structure in materials such as hydrogels or other biopolymers. The perfusion around the in vitro cell assembly shall mimic the conditions in the in vivo heart tissue, which requires that the microfluidic conditions in the HoC are precisely designed and dimensioned.

Table 5 lists commonly expressed user needs related to drug testing on the design in line with the cellular prerequisites highlighted above, together with several others for technical, analytical, and management functions. Also, a few business-related needs are mentioned, although the conceptual design is not at this stage aiming at a market analysis of a HoC; focus is on utility and technical performance. However, manufacturing and marketing must inevitable sooner or later be addressed in the product development process, in which the design alternatives may play a decisive role in manufacturing, marketing, and price.

Table 5. User needs of a HoC micro-bioreactor with cardiac cells intended for compound testing.

User Needs	Target Metrics	Specification
Biological needs		
Co-culture of cardiomyocytes/fibroblasts	Number other cell types than CM	2–4 other cell types
Cardiomyocyte assemblies beating	Beats per minutes (bpm)	30–100
cardiac cells clustered in aggregate	Number of cells per aggregate	500–1000
Sufficient cells in HoC to generate measurable signals	Cardiac cells per HoC chamber	500,000–1,000,000
Extracellular matrix created inside MBR	Type of biomaterials	PEG, Matrigel
Technical needs		
Shear force on cells corresponds in vivo of liquid media (nutrients, test solutions)	Distribution of flow rates in HoC Psi/cm	±10%
Thermostable condition for cells in HoC	Temperature range inside HoC	35–38 °C
Sampling ports for HoC effluent fluid	No. of ports and where	in: 2–3, out: 1
Oxygen transfer to cardiac bodies	Dissolved oxygen tension in aggregates	above 5%
Non-toxic fabrication materials of MBR	Type of materials	Plastics, metal
Sterile conditions	Sterility time	2 weeks
Information needs		
In situ non-destructive inspection of cells	Methods; performance	Confocal microscopy Magnification ×50
In situ observation of biomarkers		
Measurement acquisition online	Methods; performance	HCI
Inline monitoring of excreted substances	Methods; performance	MS, immunosensor
Product and manufacturing requirements		
Production cost per device/10,000 per year	EUR/unit	2–4 EUR
Consumable cost per assay	Range EUR/assay	1–5 EUR
Technician training time	Days	3 days

Clearly, there are many similarities between the user needs for a HoC device shown in Table 5 and the previous MBR design for process development (Table 4). However, the requirements on the biological systems are in the HoC more demanding due to the complex tissue architecture of the cardiac tissue. The minimal tissue equivalent, i.e., the smallest number of cells that can recapitulate the active in vivo heart tissue unit, which must be defined and realized in the HoC chambers, is a key issue in the prototype design. Also, supportive cells, e.g., fibroblasts and nerve cells for synchronized contraction (Purkinje cells), and the structure of the myocardial extracellular matrix (ECM) should be a part of that equivalent. The interaction with the culture media plays, in this respect, a critical role in maintaining the cells' responses. The technical system functions as described in the previous MBR example remain largely unchanged for a HoC prototype, but are required to be extended with electrical or optical measurement functions for recording the contractions of the cell clusters. The information functions should, in the HoC, include methods to analyze the motility of the cells or cell clusters.

These user requirements suggest a Hubka-Eder map with function systems as depicted in Figure 8. The biological system functions ($\sum$ BioSystems) have additional subsystems for building up the heart tissue units. The other $\sum$ Systems remain much the same as in the previous example. The interactions between the $\sum$ BioSystems and the other systems will, in this application, show a higher degree of complexity due to the three cell types and the myocardial ECM. The systems' interactions with the three phases of the transformation process are in some parts similar, in others different. The preparation phase has a key role in establishing the tissue structure to generate contraction between cardiomyocytes.

The sensors for monitoring the cell aggregates in situ are critical for generating the information that the computational procedures in the management system shall process further into interpretable data.

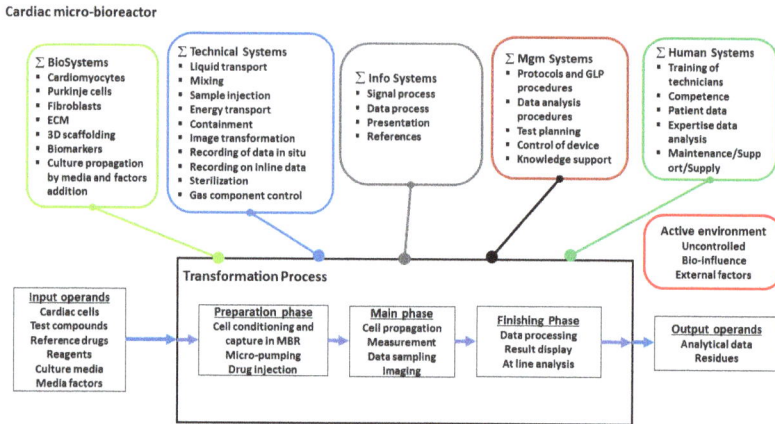

Figure 8. Hubka-Eder map showing the functional systems and subsystems that could be involved in a heart-on-a-chip MBR for purposes indicated in Table 5.

From the HE-map, the functional components related to the biological subsystems, i.e., cells and tissue functions, are identified (Figure 9A). These components can be arranged in a multitude of configurations (Figure 6B shows two possibilities of configurations). The upper configuration (I) is a design in which all cell types are confined in one body and these bodies are contained in the chip. The lower configuration (II) is a design based on parallel containment units in which cardiac bodies are kept apart. Additional configurations are, in this HoC application, more motivated to generate due to the extended complexity of the biological systems (not shown here) When comparing the two configurations in Figure 6 with the specification results in a ranking (with $w = 1.0$), which favors Configuration II when using the same weights for all need attributes (Figure 6C). Interestingly, if the weights are differentiated according to the importance of the need attributes, the ranking order remains.

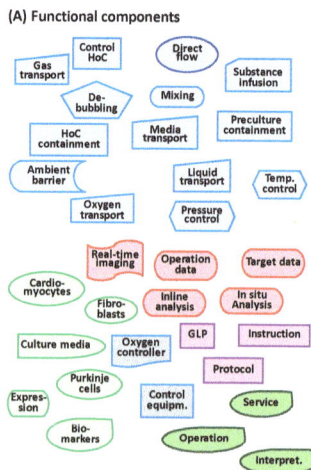

Figure 9. *Cont.*

(B) Configurations

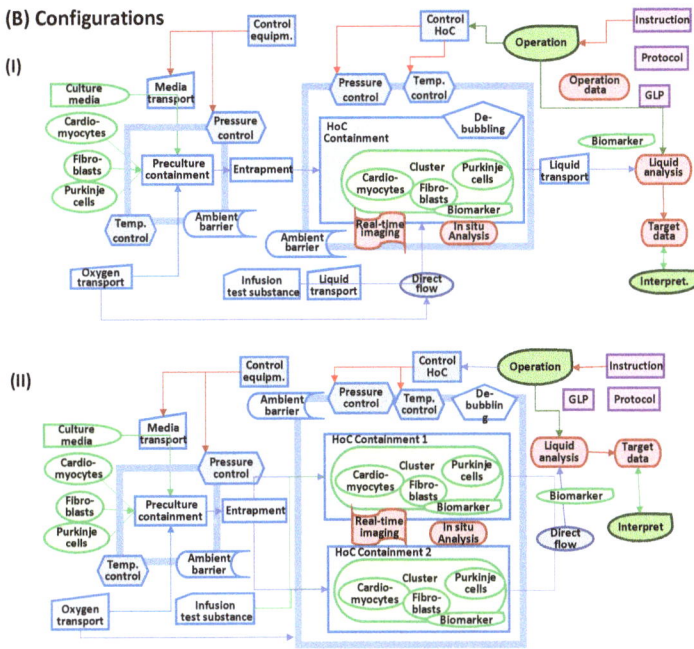

(C) Scoring of user needs (HoC for drug testing)

	User needs	Configuration I	Configuration II	Weights
1	HoC device allows co-cultures of cardiac cells	5	5	50%
2	HoC device shall remain sterile for at least 2 weeks experimentation	5	4	50%
3	Clusters of cardiac cells shall be confined in HoC device chamber	5	5	80%
4	Number of cells in HoC should correspond to *in vivo* heart tissue equivalent	5	5	80%
5	HoC shall allow continuous perfusion of culture media	5	5	80%
6	HoC should have as confocal imaging optics and inline measurement possibilities for monitoring beating rate and troponin release	2	4	80%
7	HoC shall be easy to scale-out for high-throughput testing to at least ten units	3	5	100%
8	HoC device shall be convenient to operate for lab technicians	2	4	50%
9	Fabrication cost of HoC shall be as low as possible	2	5	200%
	Score with 100% weights	34	42	
	With weight factor as in column	*26.6*	*34.7*	

Figure 9. (A) Functional components required for a design of a MBR with cardiac cells based on the functional subsystems in Figure 7. (**B**) Two possible configurations of these functional components. (**C**) A comparison of the two configurations with the user needs in Table 4 in which scores are calculated for weight factors (all of which are 100%) or as shown in last column in the table.

Next, the chosen conceptual design alternative (Configuration II) is translated into a blueprint of a heart-on-a-chip prototype in which a selection of real physical components is configured. Table S2 (Supplementary Materials) [82–87] shows the ranking of real components/devices and how these components are compared and scored versus the same need attributes that originally were identified by the users. The total scoring guides the team in the configuration of the HoC prototype using the components with the highest scores. Figure 10 shows an example of a blueprint of a previously published HoC device [79] with components fabricated in PDMS material using soft lithography, with a syringe pump, with an in situ microscopy, and with software analyzing the beating pattern of the cardiac cell clusters. The device is based on the guidance from the conceptual methodology. An alternative HoC design using commercial plastic components is described elsewhere in this issue [84].

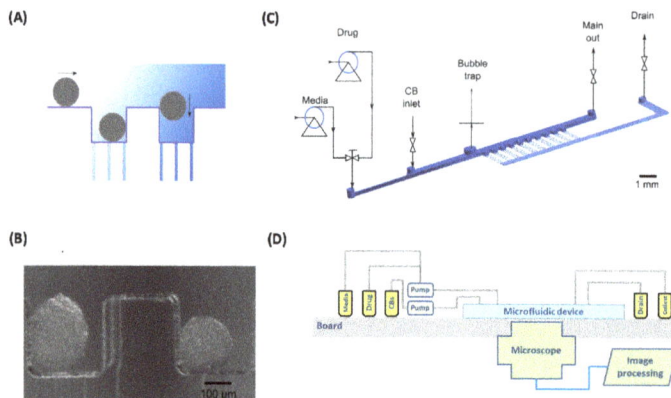

Figure 10. A blueprint of a Heart-on-Chip device designed based on the HE-map conceptual analysis in Figures 8 and 9 and the scoring of components in Table S2 (from Bergström et al. [79]). (**A**) Micro-wells with cardiac bodies; (**B**) micrograph showing two beating cardiac bodies in two wells; (**C**) graphic of the microfluidic device, and (**D**) the HoC setup.

5. Conclusions and Outlook

Although the complexity of the designed products varies, the same conceptual design methodology can be applied when designing MBRs and OoCs. The two MBR examples discussed in this review—(1) an MBR prototype for process development of recombinant mammalian cells for monoclonal antibody production and (2) an HoC prototype for drug testing of cardiac cells—elucidate this at two levels of complexity of biological systems.

Other OoC devices with higher degrees of complexity and multi-compartmental organization, such as brain tissues (e.g., blood-brain-barrier (BBB) chips) and the human eye (eye-on-a-chip) [88,89], exhibit design challenges beyond what is shown here for the HoC. For example, to recapitulate the membrane structure of blood capillaries in the brain or to recapitulate the functions of the multi-layer cellular assemblies of the cornea [29] and retina [90] or the blinking function of the eyelid [91] requires elaborate design efforts. Especially, the integration with the surrounding biopolymers (ECMs and membranes) forming scaffolds for realizing the 3D architecture of the organ then becomes an essential function in the design. Design work with the BBB- and eye-on-a-chip devices could benefit from the structural approach of conceptual design methodology.

One of the most important issues in the MBR design is to successfully mimic reality. For the heart-on-a-chip case, as well as other OoCs, that is about mimicking conditions in vivo of a human tissue. For the microbioreactor with a recombinant mammalian cell culture producing monoclonal antibodies, it is to mimic the conditions of a large-scale bioreactor with the same culture in the MBR. The ability of the MBR prototype to do this must be assessed experimentally by comparing data derived from the MBRs with data from the real system, i.e., the in vivo heart tissue or the large-scale bioreactor. Only when these data coincide sufficiently well, the prototype design can be considered successful. The bioanalytical possibilities to verify the correlation between the real system and the MBRs are considerably more demanding for the in vivo systems, e.g., the human heart tissue in a patient, and for some parameters this is not even possible in practice. For the large-scale bioreactor this is comparatively easy. Thus, the outcome of the design of a HoC prototype may not be satisfactorily assessed. This will most probably be the case for the mammalian cell culture MBR.

The two design examples in this review exhibit modest cellular complexity. This is the case especially for the MBR with the recombinant mammalian cells, which is a homogeneous mono-cellular culture; the cardiac chip is more complex, as it has a multi-cellular structure, but is limited to three cell types (cardiomyocytes, fibroblasts, and nerve cells). Still, the network of interactions between functions and real components in the design becomes significant and demanding to comprehend for the design team. This justifies the conceptual methodology with its systematic evaluation of design alternatives. The conceptual approach contributes to reducing unsuccessful prototyping of devices and facilitates the finding of favorable designs for design purposes.

Other supporting methods, such as mathematical modeling, should, however, not be neglected as useful tools to support the conceptual design. Mathematical models and dynamic simulations can be very useful, e.g., for estimating rates and dimensions in the cellular system of the organ [92]. Also, computer-aided design combined with conceptual design has proved useful in bioengineering, for example, with stem cell-derived cardiomyocytes [93]. Thus, other established engineering design methods should therefore support and complement conceptual design methods, in parallel or initiated by the conceptual directions.

Supplementary Materials: The following are available online at http://www.mdpi.com/2306-5354/5/3/56/s1, Table S1: Scoring of user needs versus real components for MBR for process development Configuration I, Table S2: Scoring of user needs versus real components for Heart-on-a-Chip Configuration I.

Funding: This research received no external funding.

Acknowledgments: The author thanks Mats Björkman, Jonas Christoffersson, and Danny van Noort for valuable discussions related to this work.

Conflicts of Interest: The author declares no conflict of interest.

References

1. Bhatia, S.N.; Ingber, D.E. Microfluidic organs-on-chips. *Nat. Biotechnol.* **2014**, *32*, 760–772. [CrossRef] [PubMed]
2. Lattermann, C.; Buchs, J. Design and operation of microbioreactor systems for screening and process development. In *Bioreactors: Design, Operation and Novel Applications*; Mandenius, C.F., Ed.; Wiley-VCH: Weinheim, Germany, 2016; pp. 35–78, ISBN 978-3-527-33768-2.
3. Huh, D.; Matthews, B.D.; Mammoto, A.; Montoya-Zavala, M.; Hsin, H.Y.; Ingber, D.E. Reconstituting organ-level lung functions on a chip. *Science* **2010**, *328*, 1662–1668. [CrossRef] [PubMed]
4. Hegab, M.H.; ElMekawy, A.; Stakenborg, T. Review of microfluidic microbioreactor technology for high-throughput submerged microbiology cultivation. *Biomicrofluidics* **2013**, *7*, 021502. [CrossRef] [PubMed]
5. Van Duinen, V.; Treetsch, J.; Joore, J.; Vulto, P.; Hankemeier, T. Microfluidic 3D cell culture: From tools to tissue models. *Curr. Opin. Biotechnol.* **2015**, *35*, 118–126. [CrossRef] [PubMed]
6. Sun, W.; Chen, Y.-Q.; Luo, G.-A.; Zhang, M.; Zhang, H.-Y.; Wang, Y.-R.; Hu, P. Organ-on-chips and its applications. *Chin. J. Anal. Chem.* **2016**, *44*, 533–541. [CrossRef]
7. Selden, C.; Fuller, B. Role of bioreactor technology in tissue engineering for clinical use and therapeutic target design. *Bioengineering* **2018**, *5*, 32. [CrossRef] [PubMed]
8. Neuzi, P.; Giselbrecht, S.; Lange, K.; Huang, T.J.; Manz, A. Revisiting lab-on-a-chip technology for drug discovery. *Nat. Rev. Drug Discov.* **2012**, *11*, 620–632. [CrossRef] [PubMed]
9. Esch, E.W.; Bahinski, A.; Huh, D. Organs-on-chips at the frontiers of drug discovery. *Nat. Rev. Drug Discov.* **2015**, *14*, 248–260. [CrossRef] [PubMed]
10. Skardal, A.; Shupe, T.; Atala, A. Organoid-on-a-chip and body-on-a-chip systems for drug screening and disease modelling. *Drug Discov. Today* **2016**, *21*, 1399–1411. [CrossRef] [PubMed]
11. Hemmerich, J.; Noack, S.; Wiechert, W.; Oldiges, M. Microbioreactor systems for accelerated bioprocess development. *Biotechnol. J.* **2018**, *13*, 1700141. [CrossRef] [PubMed]
12. Schapper, D.; Zainal Alam, M.N.H.; Szita, N.; Eliasson Lantz, A.; Gernaey, K.V. Application of microbioreactors in fermentation process development: A review. *Anal. Bioanal. Chem.* **2009**, *395*, 679–695. [CrossRef] [PubMed]
13. Betts, J.I.; Baganz, F. Miniature bioreactors, current practice and future opportunities. *Microb. Cell Fact.* **2006**, *5*, 21. [CrossRef] [PubMed]
14. Duertz, W.A. Microtiter plates as mini-bioreactors: Miniaturization of fermentation methods. *Trends Microbiol.* **2007**, *15*, 469–475. [CrossRef] [PubMed]
15. Van Noort, D. Bioreactors on a chip. In *Bioreactors: Design, Operation and Novel Applications*; Mandenius, C.F., Ed.; Wiley-VCH: Weinheim, Germany, 2016; pp. 77–112, ISBN 978-3-527-33768-2.

16. Zanzotto, A.; Boccazzi, P.; Lessard, P.; Sinskey, A.J.; Jensen, K.F. Membrane-areated microbioreactor for high-throughput bioprocessing. *Biotechnol. Bioeng.* **2004**, *87*, 243–254. [CrossRef] [PubMed]
17. Zhang, Z.; Perozziello, G.; Boccazzi, P.; Geschke, O.; Sinskey, A.J.; Jensen, K.F. Microbioreactor for bioprocess development. *J. Assoc. Lab. Autom.* **2007**, *12*, 143–151. [CrossRef]
18. Rameez, S.; Mostafa, S.S.; Miller, C.; Shukla, A.A. High-throughput miniaturized bioreactors for cell culture process, development: Reproducibility, scalability, and control. *Biotechnol. Prog.* **2014**, *30*, 718–727. [CrossRef] [PubMed]
19. ambr®250. Available online: https://www.sartorius.com/sartorius/en/EUR/products/bioreactors-fermentors/single-use/ambr-250 (accessed on 1 July 2018).
20. BioLector. Available online: https://www.m2p-labs.com/bioreactors/ (accessed on 1 July 2018).
21. Khetani, S.R.; Bhatia, S.N. Microscale culture of human liver cells for drug development. *Nat. Biotechnol.* **2008**, *26*, 120–126. [CrossRef] [PubMed]
22. Polini, A.; Prodanov, L.; Bhise, N.S.; Manoharan, V.; Dokmeci, M.R.; Khademhosseini, A. Organs on a chip a new tool for drug discovery. *Exp. Opin. Drug Discov.* **2014**, *9*, 335–352. [CrossRef] [PubMed]
23. Freyer, N.; Greuel, S.; Knöspel, F.; Gerstmann, F.; Storoch, L.; Damm, G.; Seehofer, D.; Foster Harris, J.; Iyer, R.; Schubert, F.; et al. Microscale 3D liver construct for hepatotoxicity testing in a perfused human in vitro culture platform. *Bioengineering* **2018**, *5*, 24. [CrossRef] [PubMed]
24. Wrzesinski, K.; Fey, S.J. Metabolic reprogramming and the recovery of physiological functionality in 3D cultures in micro-bioreactors. *Bioengineering* **2018**, *5*, 22. [CrossRef] [PubMed]
25. Mandenius, C.F.; Andersson, T.B.; Alves, P.M.; Batzl-Hartmann, C.; Björquist, P.; Carrondo, M.J.T.; Chesne, C.; Coecke, S.; Edsbagge, J.; Fredriksson, J.M.; et al. Towards preclinical predictive drug testing for metabolism and hepatotoxicity by in vitro models derived from human embryonic stem cells: A report on the Vitrocellomics EU-project. *Altern. Lab. Anim.* **2011**, *39*, 147–171. [PubMed]
26. Tajsoleiman, T.; Abdekhodaie, M.J.; Gernaey, K.V.; Krühne, U. Efficient computational design of a cartilage cell regeneration. *Bioengineering* **2018**, *5*, 33. [CrossRef] [PubMed]
27. Seo, J.; Huh, D. A human blinking eye-on-a-chip. In Proceedings of the 18th International Conference on Miniaturized Systems for Chemistry and Life Sciences, San Antonio, TX, USA, 26–30 October 2014.
28. Rountree, C.M.; Raghunathan, A.; Troy, J.B.; Saggere, L. Prototype chemical synapse chip for spatially patterned neurotransmitter stimulation of the retina ex vivo. *Microsyst. Nanoeng.* **2017**, *3*, 17052. [CrossRef]
29. Puleo, C.M.; McIntosh Ambrose, W.; Takezawa, T.; Elisseeff, J.; Wang, T.H. Integration and application of vitrified collagen in multi-layered microfluidic devices for corneal micro-tissue culture. *Lab Chip* **2009**, *9*, 3221–3227. [CrossRef] [PubMed]
30. Materne, E.M.; Maschmeyer, I.; Lorenz, A.K.; Horland, R.; Schimek, K.M.; Busek, M.; Sonntag, F.; Lauster, R.; Marx, U. The multi-organ chip—A microfluidic platform for long-term multi-tissue coculture. *J. Visual. Exp.* **2015**, *98*, e52526. [CrossRef] [PubMed]
31. Bauer, S.; Huldt, C.W.; Kanebratt, K.P.; Durieux, I.; Gunne, D.; Andersson, S.; Ewart, L.; Haynes, W.G.; Maschmeyer, I.; Winter, A.; et al. Functional coupling of human pancreatic islets and liver spheroids on-a-chip: Towards a novel human ex vivo type 2 diabetes model. *Sci. Rep.* **2017**, *7*, 14620. [CrossRef] [PubMed]
32. Toh, Y.-C.; Raja, A.; Yu, H.; van Noort, D. A 3D microfluidic model to recapitulate cancer cell migration and invasion. *Bioengineering* **2018**, *5*, 29. [CrossRef] [PubMed]
33. Kühlbach, C.; da Luz, S.; Mueller, M.M.; Baganz, F.; Volker, C.; Hass, V.C. A microfluidic system for investigation of tumor cell extravasation. *Bioengineering* **2018**, *5*, 40. [CrossRef] [PubMed]
34. Gernaey, K.V.; Baganz, F.; Franco-Lara, E.; Kensy, F.; Krühne, U.; Luebberstedt, M.; Marx, U.; Palmquist, E.; Schmid, A.; Schubert, F.; et al. Monitoring and control of microbioreactors: An expert opinion on development needs. *Biotechnol. J.* **2012**, *7*, 1308–1314. [CrossRef] [PubMed]
35. Maoz, B.M.; Herland, A.; Henry, O.Y.F.; Leineweber, W.D.; Yadid, M.; Doyle, J.; Mannix, R.; Kujala, W.D.; FitzGerald, E.A.; Parker, K.K.; et al. Organs-on-Chips with combined multi-electrode array and transepithelial electrical resistance measurement capabilities. *Lab Chip* **2017**, *17*, 2294–2302. [CrossRef] [PubMed]
36. Zhang, Y.S.; Aleman, J.; Shin, S.R.; Kilic, T.; Kim, D.; Shaegh, S.A.; Massa, S.; Riahi, R.; Chae, S.; Hu, N.; et al. Multi-sensor integrated organs-on-chips platform for automated and continual in situ monitoring of organoid behaviors. *Proc. Natl. Acad. Sci. USA* **2017**, *114*, E2293–E2302. [CrossRef] [PubMed]

37. Fernandez, A.C.; Halder, J.M.; Nestl, B.M.; Hauer, B.; Gernaey, K.V.; Krühne, U. Biocatalyst screening with a twist: Application of oxygen sensors integrated in microchannels for screening whole cell biocatalyst variants. *Bioengineering* **2018**, *5*, 30. [CrossRef] [PubMed]

38. Hubka, V.; Eder, E.W. *Design Science: Introduction to the Needs, Scope and Organization of Engineering Design Knowledge*, 1st ed.; Springer: Berlin, Germany, 1996.

39. Pahl, G.; Beitz, W.; Feldhusen, J.; Grote, K. *Engineering Design: A Systematic Approach*, 3rd ed.; Springer: Berlin, Germany, 2007.

40. Ulrich, K.T.; Eppinger, S.D. *Product Design and Development*, 3rd ed.; McGraw-Hill: New York, NY, USA, 2007.

41. Derelöv, M.; Jonas Detterfelt, J.; Mats Björkman, M.; Mandenius, C.-F. Engineering design methodology for bio-mechatronic products. *Biotechnol. Prog.* **2008**, *24*, 232–244. [CrossRef] [PubMed]

42. Mandenius, C.-F.; Björkman, M. Mechatronics design principles for biotechnology product development. *Trends Biotechnol.* **2010**, *28*, 230–236. [CrossRef] [PubMed]

43. Mandenius, C.F.; Björkman, M. Scale-up of bioreactors using biomechantronic design methodology. *Biotechnol. J.* **2012**, *7*, 1026–1039. [CrossRef] [PubMed]

44. Mandenius, C.-F. Biomechatronics for designing bioprocess monitoring and control systems: Application to stem cell production. *J. Biotechnol.* **2012**, *162*, 430–440. [CrossRef] [PubMed]

45. Mandenius, C.F. Design of monitoring and sensor systems for bioprocesses using biomechatronic principles. *Chem. Eng. Technol.* **2012**, *35*, 1412–1420. [CrossRef]

46. Gerlach, I.; Hass, V.C.; Mandenius, C.F. Conceptual design of an operating training simulator for a bio-ethanol plant. *Processes* **2015**, *3*, 664–683. [CrossRef]

47. Christoffersson, J.; van Noort, D.; Mandenius, C.-F. Developing organ-on-a-chip concepts using bio-mechatronic design methodology. *Biofabrication* **2017**, *9*, 025023. [CrossRef] [PubMed]

48. Mandenius, C.-F.; Björkman, M. *Biomechatronic Design in Biotechnology—A Methodology for Development of Biotechnological Products*, 1st ed.; Wiley & Sons: New York, NY, USA, 2011.

49. Yetisen, A.K.; Volpatti, L.R.; Coskun, A.F.; Cho, S.; Kamrani, E.; Butt, H.; Khademhosseinidfgh, A.; Yun, S.H. Entrepreneurship. *Lab Chip* **2015**, *15*, 3638–3660. [CrossRef] [PubMed]

50. Junaid, A.; Mashaghi, A.; Henkemeier, T.; Vulto, P. An end-user perspective on Organ-on-a-chip: Assays and usability aspects. *Curr. Opin. Biomed. Eng.* **2017**, *1*, 15–22. [CrossRef]

51. Nienow, A.W.; Rielly, C.D.; Brosnan, K.; Bargh, N.; Lee, K.; Coopman, K.; Hewitt, C.J. The physical characterisation of a microscale parallel bioreactor platform with an industrial CHO cell line expressing an IgG4. *Biochem. Eng. J.* **2013**, *76*, 25–36. [CrossRef]

52. Janakiraman, V.; Kwiatkowski, C.; Kshirsagar, R.; Ryll, T.; Huang, Y.M. Application of high-throughput mini-bioreactor system for systematic scale-down modeling, process characterization, and control strategy development. *Biotechnol. Prog.* **2015**, *31*, 623–632. [CrossRef] [PubMed]

53. Hsu, W.T.; Aulakh, R.P.S.; Traul, D.L.; Yuk, I.H. Advanced microscale bioreactor system: A representative scale-down model for bench-top bioreactors. *Cytotechnology* **2012**, *64*, 667–678. [CrossRef] [PubMed]

54. Zhang, Z.; Boccazzi, P.; Choi, H.G.; Perozziello, G.; Sinskey, A.J.; Jensen, K.F. Microchemostat, a microbial continuous culture in a polymer-based, instrumented microbioreactor. *Lab Chip* **2006**, *6*, 906–913. [CrossRef] [PubMed]

55. Wurm, F.M. Production of recombinant protein therapeutics in cultivated mammalian cells. *Nat. Biotechnol.* **2004**, *22*, 1393–1398. [CrossRef] [PubMed]

56. Huang, Y.M.; Hu, W.W.; Rustandi, E.; Chang, K.; Yusuf-Makagianser, H.; Ryll, T. Maximizing productivity of CHO cell-based fed-batch culture using chemically defined media conditions and typical manufacturing equipment. *Biotechnol. Prog.* **2010**, *26*, 1400–1410. [CrossRef] [PubMed]

57. Griffiths, J.B. Mammalian cell culture reactors, scale-up. In *Encyclopedia of Industrial Biotechnology*; Flickinger, M.C., Ed.; John Wiley & Sons: Hoboken, NJ, USA, 2010; Volume 5, pp. 3228–3241.

58. Nienow, A.W. Reactor engineering in large scale animal cell culture. *Cytotechnology* **2006**, *50*, 9–33. [CrossRef] [PubMed]

59. Becker, E.; Florin, L.; Pfizenmaier, K.; Kaufmann, H. An XBP-1 dependent bottle-neck in production of IgG subtype antibodies in chemically defined serum-free Chinese hamster ovary (CHO) fed-batch processes. *J. Biotechnol.* **2008**, *135*, 217–223. [CrossRef] [PubMed]

60. Xu, P.; Clark, C.; Ryder, T.; Sparks, C.; Zhou, J.; Wang, M.; Russell, R.; Scott, C. Characterization of TAP Ambr 250 disposable bioreactors, as a reliable scale-down model for biologics process development. *Biotechnol. Prog.* **2017**, *33*, 478–789. [CrossRef] [PubMed]

61. Kommenhoek, E.E.; van Leeuwen, M.; Gardeniers, H.; van Gulik, W.M.; van den Berg, A.; Li, X.; Ottens, M.; van der Wielen, L.A.M.; Heijnen, J.J. Lab-scale fermentation tests of microchip with integrated electrochemical sensors for pH, temperature, dissolved oxygen and viable biomass concentration. *Biotechnol. Bioeng.* **2008**, *99*, 884–892. [CrossRef] [PubMed]

62. Thuenauer, R.; Juhasz, K.; Mayr, R.; Fruhwirth, T.; Lipp, A.M.; Balogi, Z.; Sonnleitner, A. A PDMS-based biochip with integrated sub-micrometre position control for TIRF microscopy of the apical cell membrane. *Lab Chip* **2011**, *11*, 3064–3071. [CrossRef] [PubMed]

63. Mandenius, C.F. *Bioreactors: Design, Operation and Novel Applications*; Mandenius, C.F., Ed.; Wiley-VCH: Weinheim, Germany, 2016; ISBN 978-3-527-33768-2.

64. LabSmith. Available online: http://labsmith.com/ (accessed on 1 July 2018).

65. Bengtsson, K.; Christoffersson, J.; Mandenius, C.F.; Robinson, N.D. A clip-on electroosmotic pump for oscillating flow in microfluidic cell culture devices. *Microfluid. Nanofluid.* **2018**, *22*, 27. [CrossRef]

66. Joeris, K.; Frerichs, J.G.; Konstantinov, K.; Scheper, T. In-situ microscopy: Online process monitoring of mammalian cell cultures. *Cytotechnology* **2002**, *38*, 129–134. [CrossRef] [PubMed]

67. Noll, T.; Biselli, M. Dielectric spectroscopy in the cultivation of suspended and immobilized hybridoma cells. *J. Biotechnol.* **1998**, *63*, 187–198. [CrossRef]

68. Tanaka, Y.; Sato, K.; Shimizu, T.; Yamoto, M.; Okano, T.; Kitamori, T. A micro-spherical heart pump powered by cultured cardiomyocytes. *Lab Chip* **2007**, *7*, 207–212. [CrossRef] [PubMed]

69. Grosberg, A.; Alford, P.W.; McCain, M.L.; Parker, K.K. Ensembles of engineered cardiac tissues for physiological and pharmacological study: Heart on a chip. *Lab Chip* **2011**, *11*, 4165–4173. [CrossRef] [PubMed]

70. Martewicz, S.; Michielin, F.; Serena, E.; Zambon, A.; Mongillo, M.; Elvassore, N. Reversible alteration of calcium dynamics in cardiomyocytes during acute hypoxia transient in a microfluidic platform. *Integr. Biol.* **2012**, *4*, 153–164. [CrossRef] [PubMed]

71. Agarwal, A.; Goss, J.A.; Cho, A.; McCain, M.L.; Parker, K.K. Microfluidic heart on a chip for higher throughput pharmacological studies. *Lab Chip* **2013**, *13*, 3599–3608. [CrossRef] [PubMed]

72. Annabi, N.; Selimovic, S.; Acevedo Cox, J.P.; Ribas, J.; Afshar Bakooshli, M.; Heintze, D.; Weiss, A.S.; Cropek, D.; Khademhosseini, A. Hydrogel-coated microfluidic channels for cardiomyocyte culture. *Lab Chip* **2013**, *13*, 3569–3577. [CrossRef] [PubMed]

73. Ren, L.; Liu, W.; Wang, Y.; Wang, J.C.; Tu, Q.; Xu, J.; Liu, R.; Shen, S.F.; Wang, J. Investigation of Hypoxia-Induced Myocardial Injury Dynamics in a Tissue Interface Mimicking Microfluidic Device. *Anal. Chem.* **2013**, *85*, 235–244. [CrossRef] [PubMed]

74. Kaneko, T.; Nomura, F.; Hamada, T.; Abe, Y.; Takamori, H.; Sakakura, T.; Takasuna, K.; Sanbuissho, A.; Hyllner, J.; Sartipy, P.; et al. On-chip in vitro cell-network pre-clinical cardiac toxicity using spatiotemporal human cardiomyocyte measurement on a chip. *Sci Rep.* **2014**, *4*, 4670. [CrossRef] [PubMed]

75. Marsano, A.; Conficconi, C.; Lemme, M.; Occhetta, P.; Gaudiello, E.; Votta, E.; Cerino, G.; Redaelli, A.; Rasponi, M. Beating heart on a chip: A novel microfluidic platform to generate functional 3D cardiac microtissues. *Lab Chip* **2016**, *16*, 599–610. [CrossRef] [PubMed]

76. Zhang, Y.S.; Arneri, A.; Bersini, S.; Shin, S.R.; Zhu, K.; Goli-Malekabadi, Z.; Aleman, J.; Colosi, C.; Busignani, F.; Dell'Erba, V.; et al. Bioprinting 3D microfibrous scaffolds for engineering endothelialized myocardium and heart-on-a-chip. *Biomaterials* **2016**, *110*, 45–59. [CrossRef] [PubMed]

77. Zweigerdt, R.; Gruh, I.; Martin, U. Your heart on a chip: IPSC-based modeling of Barth-syndrome-associated cardiomyopathy. *Cell Stem Cell* **2014**, *15*, 9–11. [CrossRef] [PubMed]

78. Jastrzebska, E.; Tomecka, E.; Jesion, I. Heart-on-a-chip based on stem cell biology. *Biosens. Bioelectron.* **2016**, *75*, 67–81. [CrossRef] [PubMed]

79. Bergström, G.; Christoffersson, J.; Zweigerdt, R.; Schwanke, K.; Mandenius, C.F. Stem cell derived cardiac bodies in a microfluidic device for toxicity testing by beating frequency imaging. *Lab Chip* **2015**, *15*, 3242–3249. [CrossRef] [PubMed]

80. iCell®Cardiomyocytes. Available online: https://cellulardynamics.com/products-services/icell-products/icell-cardiomyocytes/ (accessed on 1 July 2018).

81. Morrison, M.; Klein, C.; Clemann, N.; Collier, D.A.; Hardy, J.; Heißerer, B.; Cader, M.Z.; Graf, M.; Kaye, J. StemBANCC: Governing access to material and data in a large stem cell research consortium. *Stem Cell Rev. Rep.* **2015**, *11*, 681–687. [CrossRef] [PubMed]
82. Mandenius, C.F.; Steel, D.; Noor, F.; Meyer, T.; Heinzle, E.; Asp, J.; Arain, S.; Kraushaar, U.; Bremer, S.; Class, R.; et al. Cardiotoxicity testing using pluripotent stem cell derived human cardiomyocytes and state-of-the-art bioanalytics: A review. *J. Appl. Toxicol.* **2011**, *31*, 191–205. [CrossRef] [PubMed]
83. Andersson, H.; Steel, D.; Asp, J.; Dahlenborg, K.; Jonsson, M.; Kågedal, B.; Jeppsson, A.; Lindahl, A.; Sartipy, P.; Mandenius, C.-F. Assaying cardiac biomarkers for toxicity testing using biosensing and cardiomyocytes derived from human embryonic stem cells. *J. Biotechnol.* **2010**, *150*, 175–181. [CrossRef] [PubMed]
84. Christoffersson, J.; Meier, F.; Kempf, H.; Schwanke, K.; Coffee, M.; Beilmann, M.; Zweigerdt, R.; Mandenius, C.F. A cardiac cell outgrowth assay for evaluating drug compounds using a cardiac spheroid-on-a-chip device. *Bioengineering* **2018**, *5*, 36. [CrossRef] [PubMed]
85. Bergström, G.; Nilsson, K.; Robinson, N.; Mandenius, C.F. Macroporous microcarriers for introducing cells in a microfluidic chip. *Lab Chip* **2014**, *14*, 3502–3504. [CrossRef] [PubMed]
86. PerkinElmer. Available online: www.perkinelmer.com/HighContent/Screening (accessed on 1 July 2018).
87. Mimetas. Available online: https://mimetas.com/page/products (accessed on 1 July 2018).
88. Estlack, Z.; Bennet, D.; Reid, T.; Kim, J. Microengineered biomimetic ocular models for ophthalmological drug development. *Lab Chip* **2017**, *17*, 1539–1551. [CrossRef] [PubMed]
89. Chan, Y.K.; Sy, K.H.; Wong, C.Y.; Man, P.K.; Wong, D.; Shum, H.C. In vitro modeling of emulsification of silicone oil as intraocular tamponade using micro-engineered Eye-on-a-Chip. *Investig. Ophthalmol. Vis. Sci.* **2015**, *56*, 3314–3319. [CrossRef] [PubMed]
90. Dodson, K.H.; Echevarria, F.D.; Li, D.; Sappington, R.M.; Edd, J.F. Retina-on-a-chip: A microfluidic platform for point access signaling studies. *Biomed. Microdevices* **2015**, *17*, 114. [CrossRef] [PubMed]
91. Seo, J.; Byun, W.Y.; Frank, A.; Massaro-Giordano, M.; Lee, V.; Bunya, V.Y.; Huh, D. Human blinking eye-on-a-chip. *Investig. Ophthalmol. Vis. Sci.* **2016**, *57*, 3872.
92. Grause, S.; Hsu, K.H.; Shafor, C.; Dixon, P.; Powell, K.C.; Chauhan, A. Mechanistic modelling of ophthalmic drug delivery to the anterior chamber by eye drops and contact lenses. *Adv. Colloid Interface Sci.* **2016**, *233*, 139–154.
93. Darkins, C.L.; Mandenius, C.F. Design of large-scale manufacturing of induced pluripotent stem cell derived cardiomyocytes. *Chem. Eng. Res. Des.* **2014**, *92*, 1142–1152. [CrossRef]

bioengineering

MDPI

Review

Role of Bioreactor Technology in Tissue Engineering for Clinical Use and Therapeutic Target Design

Clare Selden [1,]* and Barry Fuller [2]

[1]   Institute for Liver and Digestive Health, Division of Medicine, Faculty of Medical Sciences, University College London, Royal Free Hospital Campus, Rowland Hill Street, Hampstead, London NW3 2PF, UK

[2]   Department of Nanotechnology, Division of Surgery & Interventional Science, Faculty of Medical Sciences, University College London, London NW3 2QG, UK; b.fuller@ucl.ac.uk

*   Correspondence: c.selden@ucl.ac.uk; Tel.: +44-207-433-2854

Received: 2 March 2018; Accepted: 18 April 2018; Published: 24 April 2018

Abstract: Micro and small bioreactors are well described for use in bioprocess development in pre-production manufacture, using ultra-scale down and microfluidic methodology. However, the use of bioreactors to understand normal and pathophysiology by definition must be very different, and the constraints of the physiological environment influence such bioreactor design. This review considers the key elements necessary to enable bioreactors to address three main areas associated with biological systems. All entail recreation of the in vivo cell niche as faithfully as possible, so that they may be used to study molecular and cellular changes in normal physiology, with a view to creating tissue-engineered grafts for clinical use; understanding the pathophysiology of disease at the molecular level; defining possible therapeutic targets; and enabling appropriate pharmaceutical testing on a truly representative organoid, thus enabling better drug design, and simultaneously creating the potential to reduce the numbers of animals in research. The premise explored is that not only cellular signalling cues, but also mechano-transduction from mechanical cues, play an important role.

Keywords: mechanotransduction; tissue engineering; cell signaling; in vitro model; bioreactor

---

1. Introduction

For tissue engineering purposes, bioreactors are used in three ways: to enable, in vitro, a mimic of the state in which cells exist in vivo so as to understand normal cell and molecular physiology; to expand cells for potential clinical use, for example in gene and cell therapies, or to mimic a pathological state in order to study the pathophysiology; and to establish new therapeutic targets and test potential new treatments in a more realistic setting than simple in vitro conventional culture. Success in this area would also reduce the burden of use of animals in pharmacological testing.

There are several other uses of bioreactors, both on a micro- and larger scale; often, small- and micro-bioreactors are used in manufacturing to design new processes of production prior to full scale fabrication, and lab-on-a-chip applications. These, however, are not the subject of this review. Rather, this review will cover, in the most part, design of bioreactors that intend to address the functional mimics of an in vivo environment.

Requirements for Bioreactor Design

Recreating the natural cellular niche using bioreactors is not trivial, and all impacts on cell behaviour must be considered. For example, there are complex stimuli in vivo that a cell may be exposed to, related to biochemical or metabolic cues on the one hand (chemical stimuli) and mechanical stimuli on the other. There is a likely interaction between these signals that will impact cell performance, so that for bioreactor design it is key to fully understand normal cell behaviour at the molecular level.

This is particularly relevant when the intention is to mimic a specific pathophysiology with the intention of promoting or testing new therapies.

In short, a bioreactor design should consider in vivo tissue structure, cellular organization, and cell survival, which will in turn influence the ensuing function, so the thought processes must start with the functional requirements; one size will never fit all. Some examples from biology include the performance of blood vessels depending on their role; for example, the make-up of a vein usually delivering low pressure flow at low shear that is responsible not only for flow but for heat dissipation, compared with an artery responsible for high flows, at much higher pressures, especially close to the heart, which are designed to have thicker musculature in vessel walls and to be more elastic to deal with greater pressures and pulsatile flow; these tissue structures are often anisotropic. To model these in a bioreactor, not only the correct cell type but also the mechanical structures capable of delivering the function is necessary. Another example would be a bioreactor to mimic solid tissues without, for xample, liver and kidney, which, in contrast, are not dependent on the alignment of particular fibers for function; these are more mechanically isotropic.

The success of static culture reactors even with 3D constructs is often limited by mass transfer issues, with either a lack of nutrients to maintain the constructs or failure from a build-up of endogenous waste products. This arises because the only movement of solutes within the construct is concentration gradient-dependent and relies only on a diffusion mechanism, so that larger molecules move more slowly across a gradient than smaller molecules.

Today's bioreactors usually contain 3-dimensional constructs of cells formed from a single phenotype; co-cultures of cells of different phenotypes, e.g., epithelial and endothelial; or epithelial and fibroblastic, or indeed a mixture of several cell types aimed at recreating the in vivo niche. Mass transfer is improved by making the bioreactors dynamic, using, simply, convection; this fluid flow facilies mass transfer. Some simple examples of these mixing bioreactors achieving the dynamic state are spinner flasks or rocking or wave form bioreactors. However, these are not really mimics of any system in the body.

2. Bioreactor Designs

2.1. Perfusion Bioreactors

Perfusion reactors, in contrast, simulate the in vivo environment more closely. The more successful microbioreactors are based on perfusion systems [1,2], some with simple downward or cross flow, and others delivering a microgravity environment. The latter achieves greater mass transfer; examples include rotating wall cell culture systems and fluidised bed bioreactors. Nonetheless, the flow must be optimised: optimal perfusion leads to improved, tissue-specific expression, whilst too much can impact not only on cell proliferation, but survival and function possibly by the removal of some important paracrine factors important for cell survival [3]. Crabbe et al. [4] utilised the rotatory cell culture system (RCCS), to improve reseeding of decellularised lung tissue with lung cells (C10) and bone marrow-derived mesenchymal stromal cells (MSCs) and to determine an effect on differentiation of the recellularised construct. They demonstrated improved proliferation and decreased apoptosis in this dynamic culture, as well as evidence of differentiation of the stromal cell component; authors speculate this improvement over 2D culture is mediated by the biomechanical force resulting from fluid shear, and the increased mass transfer of nutrients, oxygen, and waste-product dilution, and suggest an application in providing engineered lung tissue and understanding the transition of normal-to-fibrotic lung phenotype, prevalent in chronic obstructive pulmonary disease (COPD) and affecting more than 60 million people.

The choice of scaffold for the tissue construct will also impact on mass transfer. The thickness of some "artificial" substrates hinders mass transfer, and pore sizes may not reflect in vivo tissue organisation. Decellularised tissues may offer a better scaffold environment.

2.2. Oxygenation

Another element that is frequently forgotten in bioreactor design is the delivery of suitable oxygen tensions, especially in bioreactors utilising culture media as the nutrient supply, since oxygen diffusion into aqueous solutions is poor, in contrast to the oxygen-carrying capacity of blood normally perfusing the body. Whilst microbioreactors can overcome this to an extent by having thin layers of liquid in the fluid path, good control of oxygen provision and consumption is difficult. Improvements in fluorescent oxygen sensors have led to advances in this area, although when the perfusion fluid has high protein content, as seen, for example, in plasma, the technology is not sufficiently robust. Oxygen delivery in whole organ bioreactors has hampered successful use [5]; for example, the metabolic demands of cardiomyocytes and hepatocytes for oxygen differ (27.6 and 18 nmol oxygen·mg protein^{-1}·min^{-1}, respectively) and are not met by a diffusional supply of oxygen in thick tissue constructs. Alternative oxygen delivery systems may require the use of perfluorocarbons [6] or more physiologically red blood cells but not whole blood, as that may introduce immune components leading to a systemic inflammatory response.

2.3. Sheer Stress

Whilst the dynamic state is favourable, since it introduces a degree of shear stress by the very nature of the flow, this also has an impact on performance. In some tissues it is advantageous, for example, in blood vessels; in others, it may not represent the physiological state, e.g., in the liver blood flow through the portal vein is 1200 mL/min; however, the metabolic cells of the liver, the hepatocytes, are protected by the sinusoidal endothelial cell fenestrae that protect the hepatocytes themselves from shear. None but the most sophisticated of bioreactors can easily mimic that.

2.4. Mechanical Stimuli

The mechanical stimuli that impact cell physiology can be engineered into bioreactors in several ways. Essentially, these stimuli are achieved by enforcing a mechanical load on a tissue or cell construct. Such forces include compression, shear stress, stretch and compression, and pressure loads. It is clear from biology that each of these are reflected in body systems: muscles, blood vessels, ligaments, and tendons are all exposed to stretch loads in different ways. Bones encounter compression and torsion in normal physiology. A broken bone has two phases of healing: that which requires no movement and that which requires a load to encourage bone and muscle growth, i.e., tissues can adapt performance according to mechanical stimulation. The two laws governing such adaptation in hard and soft tissues, respectively, were described in the 19th century and still hold true today (Wolff 1892 and Davis 1867) [7,8].

2.5. Mechano-Transduction and Cellular Signalling

At a more cellular level, these loads lead to mechano-transduction of cellular signalling pathways. Examples include focal adhesions, cell to cell contacts, integrins and cadherins, and nuclear deformation. Such changes take from seconds to weeks depending on whether they are receiving a stimulus such as surface rigidity, "sensing" the local environment (milliseconds to seconds), altering gene expression (minutes to hours), or changing cell behaviour and function (days), or even influencing tissue development (weeks). Proteins involved in these processes include focal adhesion kinases and YAP (yes-associated protein), among many others.

An area of burgeoning research is the impact of viscosity and stiffness on cellular signalling; viscosity and stiffness also impose a mechanical load and affect cell morphology [9]. This too should be encompassed in microbioreactor design. As well as mechano-transduction impacting on signalling, downstream gene expression will be altered; the role of the directional loading force can influence protein binding on extracellular matrices and thus is also critical in bioreactor design. The shear stress forces should represent the mechanical environment of the original tissue [10].

Other mechanically induced stimuli can lead to tissue differentiation; an example is the fate of stem cells subjected to a load, known as mechano-differentiation [11–14].

Stretch has most often been described in the context of vascular tissue engineering, and is effectively described as the "new" length divided by the initial length. Directional change can be in any direction, and the cyclic stretch observed in muscles leads to enhancement of protein expression and ecm protein content [15]. Often, the result of such stretch forces in bioreactor design is cell alignment that better represents the normal in vivo tissue environment. Cardiac tissues constructed under mechanical stretch and/or electrical stimuli display propagation speeds similar to those observed in vivo and respond to electrical stimuli by synchronised contractions [16,17].

2.6. Examples Used in Tissue Engineering and Pathophysiological Studies

Burk et al. [18], using decellularised tendons reseeded with mesenchymal stromal cells, applied mechanical stimulation with a cyclic-strain bioreactor. Natural horse tendon movement is best represented at a frequency of 1 Hz, which these authors used, and a strain of 2% was applied as a close estimate of that seen in the superficial digital flexor tendon in horses, and is comparable to that of the Achilles tendon in man. Stepwise time increases in strain and rest periods were implemented. Their data indicated cell anisotropy when comparing cells grown on scaffolds rather than monolayer, and increased differentiation; however, a negative impact on cell viability possibly arising from poor cell adaptation to the strain/rest cycles re-iterates the importance of the design elements of the bioreactor.

Examples of pressure in vivo are well described: atmospheric pressure on skin, i.e., a load distributed over an area, usually uniformly. Groeber et al. [19] devised a bioreactor to produce a vascularised skin construct that was comprised of 2 cell types initially in a fluid circuit in a BioVacSc. The perfusion flow produced 10 mmHg, initially rising to 80 mmHg, prior to the typical physiologic pulsatile pressure profile with systolic at 120 and diastolic 80 mmHg, respectively. Two fluidic circuits were added to deliver fluids to both the surface and underside of the construct for 6 days. Thereafter, the addition of human embryonic kidney (hEK) cells to the surface of the construct completed the model. Changes to the nutrient media and a switch to air-liquid interface conditions for 27 days revealed a well-stratified epidermis with appropriate structural layers and a strong epidermal barrier. The most important parameters were recreated in the bioreactor, and this model should enable studies on the interaction of cellular and non-cellular blood compartments with the dermal layers, which are useful for immunological research and with clinical potential for deep wounds.

2.7. Stretch/Compression

Arteries are subjected to blood pressure loads, and heart valves are subjected to alternating pressures, so that the valves receiving a pressure gradient signal respond by opening and closing. Clearly, this is a complex pattern of events, and when reproduced in bioreactors it enables tissue constructs that are closely aligned with tissues in vivo.

Egger et al. [20] demonstrated a parallel perfusion bioreactor, in which 8 conditions could be compared simultaneously under physiological shear stress and hydrostatic pressures necessary for differentiating osteogenic tissues from mesenchymal stem cells. The authors noted that when using an artificial scaffold the mechanical cues of fluid sheer stress led to sheer forces (10–40 mPa), an order of magnitude lower than observed in vivo, typically 0.3–3 Pa [21]. The porosity and pore sizes of the scaffold were responsible for the fluid sheer stresses that impacted cell and protein deposition; under dynamic flow expression of alkaline phosphatase, an estimate of osteogenic differentiation was highest, although hydrostatic pressure was not influential.

Compression is related to stretch in as much as it is directional but the opposite in force outcome: stretch leads to a greater than initial, and compression a lesser than original, dimension, as evidenced in nature by, for example, weight-bearing joints and cartilage such as the knee joint. Bioreactors mimicking compression models result in more ordered structures and increased mucopolysaccharide content of the extracellular matrix [22–24]. Even blood vessel lumens experience compression, but in a radial direction resulting from changes in blood pressure. The cellular reactions follow a force at two levels; the macroscopic force event leads to changes at the microscopic level due to the shear induced by the fluid flows from the tissue under compression.

In physiology, all of these forces may be acting together in a particular tissue, in, for example, arteries, heart muscle, and musculoskeletal tissues. Only bioreactors designed to mimic all of these forces including bending and torsion will provide a good model for studying normal and pathophysiological production of tissue constructs for clinical application and testing of potential therapeutic modalities.

2.8. Cell Seeding

The manner in which cells are seeded into bioreactors will also impact considerably their performance; in many cases, the intention is for cells to be uniformly seeded throughout the scaffold. Thus, pore size and the model of entry for cells will play an important role, not only during seeding but also for nutrient/metabolite and gas exchange during subsequent culture. Whilst simply applying a cell suspension to scaffold enables some cell attachment, and even infiltration deeper into the scaffold pores, it is an uncontrolled process and thus subject to considerable variability. Approaches that use different physical processes to drive the processes are better, such as acoustic or electromagnetics energy or even vacuum application [25–27].

2.9. In Silico Modelling

Since there is considerable complexity required in bioreactor design to mimic in vivo organ function, utilising computational modelling (in silico) from knowledge of normal physiology prior to bioreactor construction may improve outcomes. Tresoldi et al. [28] utilised a computational model with fluid-structure interaction to demonstrate the necessary parameters for vascular tissue engineering. Using the model in the MiniBreath bioreactor (pulsatile perfusion), it predicted pressures acting on the tubular scaffolds, circumferential deformation, solid components, and wall shear stresses, with good comparability with the analytical model. The model could not take account of scaffold thickness along the length, nor predict any changes associated with cell growth and maturation, thus emphasizing the importance of defining the key questions prior to modelling and appreciating the limitations of modelling versus analytical study. Nonetheless, for a complex system, much information can be gleaned form an in silico approach.

2.10. Scaffolds Used with Bioreactors

The choice of scaffold for tissue constructs, be it synthetic polymer materials (e.g., poly lactic acid, poly caprolactone, or polyglycolic acid), bioceramic-based or natural polymers (e.g., collagens, dextrans, gelatins, alginates, hyaluronic acids etc.), has the aim of aiding cell survival, function, and proliferation whilst concomitantly minimising immune responses, which is especially important for tissues to be used in clinical applications, such as heart valves. These scaffolds have been widely reviewed elsewhere [29–41], including their use with bioreactor technology. Typically, they are porous to increase cell/scaffold area and promote 3-dimensional growth of the seeded cells. A more recently exploited scaffold principle is that of using decellularised natural organs generated to maintain extracellular matrix components but be devoid of cellular DNA or protein (see following sections). Since there are no cellular components left on decellularised scaffolds, there should not be species cross reactivity, so that, for example, decellularised porcine organs could potentially be used for recellularisation with human cell material; nonetheless, the extracellular matrices of different species do indeed differ, so they may introduce incompatibilities, depending on the performance requirements. The choice of scaffold is likely to be key to producing a bioreactor system that achieves its aim, and several bioreactors incorporate such scaffolds.

The next sections will consider some case studies of bioreactors incorporating some or all of the above principles.

3. Bioreactors Used to Provide Tissue Constructs for Implantation

Ma et al. [42], utilising decellularised aortae derived from foetal pigs and seeded with canine endothelial cells for three days in a principally static environment to encourage cell attachment, were subjected to dynamic bioreactor culture for a further 7 days prior to implanting into carotid arteries. The bioreactor imposed a liquid flow starting at 20 mL/min, which gradually increased daily by 10 mL/min up to 60 mL/min; both static pressures (10 mmHg) and dynamic pressures (60 mmHg) were imposed as would occur in vivo. Six months after implantation, the grafts had remodelled, appearing as normal arteries with complete endothelial cell layers arising from those implanted, those native to the animal, and those from endothelial progenitor cells originating from the blood.

Whilst Ghaedi et al. [43] generated bioengineered lung tissue from induced pluripotent stem cell (iPSC)-derived epithelial cells on decellularised lung tissues from rat and human lungs, they emphasized the importance of good gas exchange across the lung, requiring integrity of the physical barrier expressed in vivo from both epithelial and mature endothelial cells with appropriate tight junctions and adhesive molecules. This study used only the epithelial cell component. Nonetheless, this example indicates that both within and across species decellularised scaffolds support the survival proliferation and function of airway epithelial cells.

The need to provide a shorter and more effective route for patients requiring a coronary artery bypass has led to tissue engineering efforts to produce small diameter vascular grafts of less than 6 mm diameter. This is a lengthy process; Tondreau et al. [44] designed a simple approach that reduced the time from more than 4 months to 4 weeks by starting with conventionally cultured fibroblast sheets that could be produced "offline", that were subsequently rolled onto mandrels, decellularised, and recellularised with patients' own endothelial cells in a perfusion bioreactor. Significant mechanical testing of the graft for burst pressure compliance, thickness, and suture strength retention (26 +/− 2 gf, compared with 138 +/− 50 gf) established a graft composed only of human dermal fibroblasts, reseeded with endothelial cells, which could shorten the timeline for patient treatment significantly.

An endothelial layer on biological graft matrices is considered important from the perspective of antithrombotic activity [45] and preventing graft failure. However, in artificial grafts, the pulsatile flow of a bioreactor can disrupt the endothelial cell surface under high flow conditions, as may be experienced by cusps during valve opening in the native valves, so that whilst one may endeavour to mimic the natural environment, some compromise in bioreactor designs may be required to enable adaptation of the recellularised grafts [46]. In a different organ system, the liver, again using decellularised tissue as the bioreactor scaffold, Hussein et al. defined a heparin-gelatin mixture as an antithrombotic agent prior to cell seeding that positively impacted attachment and migration of endothelial cells, as well as leading to enhanced function from the parenchymal fraction of subsequently seeded HepG2 epithelial cells [47].

Due to a shortage of organs for transplant, the research endeavour to "grow" organs for transplantation, and improvements in the decellularisation/recellularisation strategies, there is potential in this area. The bioreactors required to implement this approach must be specific for that organ's physiology (for example, for the type of flow required, whether continuous perfusion or pulsatile flow [48]).

4. Bioreactors Designed for Disease Modelling

Tumour cell modelling for testing of new drugs in human cells, equivalent to tumour biopsies, would be a significant advance. Nietzer et al. [49] produced such a bioreactor using a jejunal decellularised scaffold and colon cancer cells with fibroblasts designed to have tissues at the interface of two separate fluid circulations meeting apical and basolateral-specific culture conditions: contemporaneously modelling optimized fluid circulations in terms of ambient pressures, inlet and outlet velocities, and defining shear stress conditions. The resultant tumour-like tissue expressed beta catenin at cell borders and had a stroma positive for vimentin and cytokeratins, which is typical of colon adenocarcinomas. This model was clearly delineated from the same cells as monolayer cultures by the 5FU (5-fluorouracil) response and exemplified the treatment response in man. There are several examples in the literature.

5. Small Bioreactors to Mimic Larger Production Bioreactors as Bioartificial Organs

Bioartificial organs not intended for transplantation but for temporary replacement of function have been developed [50–53], in particular for the liver system, since the liver, being highly regenerative, can repair itself given time after an insult of acute liver failure. Whilst organ transplants are curative, the lack of donor organs results in many dying before they receive a transplant. Several experimental models exist, and a few are in clinical trials. However, from a bioreactor perspective, culturing on a human scale does not enable rapid prototyping or optimisation of production conditions. Our own group has produced a bioartificial liver machine (BAL) on human scale based on a fluidised bed bioreactor design, which maximises mass transfer (UCLBAL), and tested it in a porcine model of acute liver failure [50,52,53]. To refine certain aspects, there was a need to develop a small scale mimic on the small (~30–50 mL) (not micro) scale but to nonetheless enable metabolic, gene expression, and protein studies, and since this small fluidized bed (sFBB) bioreactor enables flow studies that can be mathematically modelled, it is more easily scalable to clinical size and enables a comparison of dynamic versus static conditions [54] suitable for drug biotransformation. Multiple units can be run simultaneously, which increases research capabilities.

6. Conclusions

Over the past decade, significant improvements in design and construction of bioreactors have been made. Systems have been developed that allow robust and reproducible culture conditions to be maintained. Specific bioreactor design is critical to the production of useful systems that can predict performance if based on a natural cell niche from in vivo physiology. Whilst the more sophisticated the bioreactor approach, the more likely it is to reflect the natural physiological state, simpler designs are likely to be more operationally robust, so a compromise based on bioreactor complexity versus the essential functional parameters of the desired end-product will always be necessary.

Author Contributions: C.S. conceived and wrote this review manuscript together with B.F.; B.F. contributed to planning the manuscript structure and writing some parts of this review.

Conflicts of Interest: The authors declare no conflict of interest.

References

1. Martin, I.; Wendt, D.; Heberer, M. The role of bioreactors in tissue engineering. *Trends Biotechnol.* **2004**, *22*, 80–86. [CrossRef] [PubMed]
2. Carrier, R.L.; Rupnick, M.; Langer, R.; Schoen, F.J.; Freed, L.E.; Vunjak-Novakovic, G. Perfusion improves tissue architecture of engineered cardiac muscle. *Tissue Eng.* **2002**, *8*, 175–188. [CrossRef] [PubMed]
3. King, J.A.; Miller, W.M. Bioreactor development for stem cell expansion and controlled differentiation. *Curr. Opin. Chem. Biol.* **2007**, *11*, 394–398. [CrossRef] [PubMed]
4. Crabbe, A.; Liu, Y.; Sarker, S.F.; Bonenfant, N.R.; Barrila, J.; Borg, Z.D.; Lee, J.J.; Weiss, D.J.; Nickerson, C.A. Recellularization of decellularized lung scaffolds is enhanced by dynamic suspension culture. *PLoS ONE* **2015**, *10*, e0126846. [CrossRef] [PubMed]
5. Kulig, K.M.; Vacanti, J.P. Hepatic tissue engineering. *Transpl. Immunol.* **2004**, *12*, 303–310. [CrossRef] [PubMed]
6. Nahmias, Y.; Berthiaume, F.; Yarmush, M.L. Integration of technologies for hepatic tissue engineering. *Adv. Biochem. Eng. Biotechnol.* **2007**, *103*, 309–329. [PubMed]
7. Wolff, J. *Das Gesetz der Transformation der Knochen*; Hirshwald: Berlin, Germany, 1892.
8. Davis, H. *Conservative Surgery*; Appleton: New York, NY, USA, 1867.
9. Gonzalez-Molina, J.; Selden, B.F.C. *Extracellular Fluid Viscosity Enhances Cell-Substrate Interaction and Impacts on Cell Size and Morphology*; eCM Meeting Abstracts 2016, Collection 5; TCES: London, UK, 2016; p. 74.

10. McFetridge, P.S.; Abe, K.; Horrocks, M.; Chaudhuri, J.B. Vascular tissue engineering: Bioreactor design considerations for extended culture of primary human vascular smooth muscle cells. *ASAIO J.* **2007**, *53*, 623–630. [CrossRef] [PubMed]

11. Lee, C.H.; Shin, H.J.; Cho, I.H.; Kang, Y.M.; Kim, I.A.; Park, K.D.; Shin, J.W. Nanofiber alignment and direction of mechanical strain affect the ecm production of human acl fibroblast. *Biomaterials* **2005**, *26*, 1261–1270. [CrossRef] [PubMed]

12. Wendt, D.; Marsano, A.; Jakob, M.; Heberer, M.; Martin, I. Oscillating perfusion of cell suspensions through three-dimensional scaffolds enhances cell seeding efficiency and uniformity. *Biotechnol. Bioeng.* **2003**, *84*, 205–214. [CrossRef] [PubMed]

13. Braccini, A.; Wendt, D.; Jaquiery, C.; Jakob, M.; Heberer, M.; Kenins, L.; Wodnar-Filipowicz, A.; Quarto, R.; Martin, I. Three-dimensional perfusion culture of human bone marrow cells and generation of osteoinductive grafts. *Stem Cells* **2005**, *23*, 1066–1072. [CrossRef] [PubMed]

14. Filipowska, J.; Reilly, G.C.; Osyczka, A.M. A single short session of media perfusion induces osteogenesis in hbmscs cultured in porous scaffolds, dependent on cell differentiation stage. *Biotechnol. Bioeng.* **2016**, *113*, 1814–1824. [CrossRef] [PubMed]

15. Nieponice, A.; Maul, T.M.; Cumer, J.M.; Soletti, L.; Vorp, D.A. Mechanical stimulation induces morphological and phenotypic changes in bone marrow-derived progenitor cells within a three-dimensional fibrin matrix. *J. Biomed. Mater. Res. Part A* **2007**, *81*, 523–530. [CrossRef] [PubMed]

16. Ott, H.C.; Matthiesen, T.S.; Goh, S.K.; Black, L.D.; Kren, S.M.; Netoff, T.I.; Taylor, D.A. Perfusion-decellularized matrix: Using nature's platform to engineer a bioartificial heart. *Nat. Med.* **2008**, *14*, 213–221. [CrossRef] [PubMed]

17. Radisic, M.; Fast, V.G.; Sharifov, O.F.; Iyer, R.K.; Park, H.; Vunjak-Novakovic, G. Optical mapping of impulse propagation in engineered cardiac tissue. *Tissue Eng. Part A* **2009**, *15*, 851–860. [CrossRef] [PubMed]

18. Burk, J.; Plenge, A.; Brehm, W.; Heller, S.; Pfeiffer, B.; Kasper, C. Induction of tenogenic differentiation mediated by extracellular tendon matrix and short-term cyclic stretching. *Stem Cells Int.* **2016**, *2016*, 7342379. [CrossRef] [PubMed]

19. Groeber, F.; Engelhardt, L.; Lange, J.; Kurdyn, S.; Schmid, F.F.; Rucker, C.; Mielke, S.; Walles, H.; Hansmann, J. A first vascularized skin equivalent as an alternative to animal experimentation. *Altex* **2016**, *33*, 415–422. [CrossRef] [PubMed]

20. Egger, D.; Spitz, S.; Fischer, M.; Handschuh, S.; Glosmann, M.; Friemert, B.; Egerbacher, M.; Kasper, C. Application of a parallelizable perfusion bioreactor for physiologic 3D cell culture. *Cells Tissues Organs* **2017**, *203*, 316–326. [CrossRef] [PubMed]

21. Weinbaum, S.; Cowin, S.C.; Zeng, Y. A model for the excitation of osteocytes by mechanical loading-induced bone fluid shear stresses. *J. Biomech.* **1994**, *27*, 339–360. [CrossRef]

22. Hung, C.T.; Mauck, R.L.; Wang, C.C.; Lima, E.G.; Ateshian, G.A. A paradigm for functional tissue engineering of articular cartilage via applied physiologic deformational loading. *Ann. Biomed. Eng.* **2004**, *32*, 35–49. [CrossRef] [PubMed]

23. Seidel, J.O.; Pei, M.; Gray, M.L.; Langer, R.; Freed, L.E.; Vunjak-Novakovic, G. Long-term culture of tissue engineered cartilage in a perfused chamber with mechanical stimulation. *Biorheology* **2004**, *41*, 445–458. [PubMed]

24. Liu, C.; Abedian, R.; Meister, R.; Haasper, C.; Hurschler, C.; Krettek, C.; von Lewinski, G.; Jagodzinski, M. Influence of perfusion and compression on the proliferation and differentiation of bone mesenchymal stromal cells seeded on polyurethane scaffolds. *Biomaterials* **2012**, *33*, 1052–1064. [CrossRef] [PubMed]

25. Van Wachem, P.B.; Stronck, J.W.; Koers-Zuideveld, R.; Dijk, F.; Wildevuur, C.R. Vacuum cell seeding: A new method for the fast application of an evenly distributed cell layer on porous vascular grafts. *Biomaterials* **1990**, *11*, 602–606. [CrossRef]

26. Ito, A.; Ino, K.; Hayashida, M.; Kobayashi, T.; Matsunuma, H.; Kagami, H.; Ueda, M.; Honda, H. Novel methodology for fabrication of tissue-engineered tubular constructs using magnetite nanoparticles and magnetic force. *Tissue Eng.* **2005**, *11*, 1553–1561. [CrossRef] [PubMed]

27. Li, H.; Friend, J.R.; Yeo, L.Y. A scaffold cell seeding method driven by surface acoustic waves. *Biomaterials* **2007**, *28*, 4098–4104. [CrossRef] [PubMed]

28. Tresoldi, C.; Bianchi, E.; Pellegata, A.F.; Dubini, G.; Mantero, S. Estimation of the physiological mechanical conditioning in vascular tissue engineering by a predictive fluid-structure interaction approach. *Comput. Methods Biomech. Biomed. Eng.* **2017**, *20*, 1077–1088. [CrossRef] [PubMed]

29. Arenas-Herrera, J.E.; Ko, I.K.; Atala, A.; Yoo, J.J. Decellularization for whole organ bioengineering. *Biomed. Mater. (Bristol Engl.)* **2013**, *8*, 014106. [CrossRef] [PubMed]

30. Hussein, K.H.; Park, K.M.; Kang, K.S.; Woo, H.M. Biocompatibility evaluation of tissue-engineered decellularized scaffolds for biomedical application. *Mater. Sci. Eng. C Mater. Biol. Appl.* **2016**, *67*, 766–778. [CrossRef] [PubMed]

31. Yu, Y.; Alkhawaji, A.; Ding, Y.; Mei, J. Decellularized scaffolds in regenerative medicine. *Oncotarget* **2016**, *7*, 58671–58683. [CrossRef] [PubMed]

32. Atias, S.; Mizrahi, S.S.; Shaco-Levy, R.; Yussim, A. Preservation of pancreatic tissue morphology, viability and energy metabolism during extended cold storage in two-layer oxygenated university of wisconsin/perfluorocarbon solution. *Isr. Med. Assoc. J.* **2008**, *10*, 273–276. [PubMed]

33. Pacifici, A.; Laino, L.; Gargari, M.; Guzzo, F.; Velandia Luz, A.; Polimeni, A.; Pacifici, L. Decellularized hydrogels in bone tissue engineering: A topical review. *Int. J. Med. Sci.* **2018**, *15*, 492–497. [CrossRef] [PubMed]

34. Chen, Y.; Chen, J.; Zhang, Z.; Lou, K.; Zhang, Q.; Wang, S.; Ni, J.; Liu, W.; Fan, S.; Lin, X. Current advances in the development of natural meniscus scaffolds: Innovative approaches to decellularization and recellularization. *Cell Tissue Res.* **2017**, *370*, 41–52. [CrossRef] [PubMed]

35. Lee, E.; Milan, A.; Urbani, L.; De Coppi, P.; Lowdell, M.W. Decellularized material as scaffolds for tissue engineering studies in long gap esophageal atresia. *Expert Opin. Biol. Ther.* **2017**, *17*, 573–584. [CrossRef] [PubMed]

36. Wang, Y.; Nicolas, C.T.; Chen, H.S.; Ross, J.J.; De Lorenzo, S.B.; Nyberg, S.L. Recent advances in decellularization and recellularization for tissue-engineered liver grafts. *Cells Tissues Organs* **2017**, *203*, 203–214. [CrossRef] [PubMed]

37. Lovati, A.B.; Bottagisio, M.; Moretti, M. Decellularized and engineered tendons as biological substitutes: A critical review. *Stem Cells Int.* **2016**, *2016*, 7276150. [CrossRef] [PubMed]

38. Nachlas, A.L.Y.; Li, S.; Davis, M.E. Developing a clinically relevant tissue engineered heart valve—A review of current approaches. *Adv. Healthc. Mater.* **2017**, *6*. [CrossRef] [PubMed]

39. Swinehart, I.T.; Badylak, S.F. Extracellular matrix bioscaffolds in tissue remodeling and morphogenesis. *Dev. Dyn. Off. Publ. Am. Assoc. Anat.* **2016**, *245*, 351–360. [CrossRef] [PubMed]

40. Boccafoschi, F.; Botta, M.; Fusaro, L.; Copes, F.; Ramella, M.; Cannas, M. Decellularized biological matrices: An interesting approach for cardiovascular tissue repair and regeneration. *J. Tissue Eng. Regen. Med.* **2017**, *11*, 1648–1657. [CrossRef] [PubMed]

41. Rana, D.; Zreiqat, H.; Benkirane-Jessel, N.; Ramakrishna, S.; Ramalingam, M. Development of decellularized scaffolds for stem cell-driven tissue engineering. *J. Tissue Eng. Regen. Med.* **2017**, *11*, 942–965. [CrossRef] [PubMed]

42. Ma, X.; He, Z.; Li, L.; Liu, G.; Li, Q.; Yang, D.; Zhang, Y.; Li, N. Development and in vivo validation of tissue-engineered, small-diameter vascular grafts from decellularized aortae of fetal pigs and canine vascular endothelial cells. *J. Cardiothoracic. Surg.* **2017**, *12*, 101. [CrossRef] [PubMed]

43. Ghaedi, M.; Le, A.V.; Hatachi, G.; Beloiartsev, A.; Rocco, K.; Sivarapatna, A.; Mendez, J.J.; Baevova, P.; Dyal, R.N.; Leiby, K.L.; et al. Bioengineered lungs generated from human ipscs-derived epithelial cells on native extracellular matrix. *J. Tissue Eng. Regen. Med.* **2017**. [CrossRef] [PubMed]

44. Tondreau, M.Y.; Laterreur, V.; Gauvin, R.; Vallieres, K.; Bourget, J.M.; Lacroix, D.; Tremblay, C.; Germain, L.; Ruel, J.; Auger, F.A. Mechanical properties of endothelialized fibroblast-derived vascular scaffolds stimulated in a bioreactor. *Acta Biomater.* **2015**, *18*, 176–185. [CrossRef] [PubMed]

45. Pompilio, G.; Rossoni, G.; Sala, A.; Polvani, G.L.; Berti, F.; Dainese, L.; Porqueddu, M.; Biglioli, P. Endothelial-dependent dynamic and antithrombotic properties of porcine aortic and pulmonary valves. *Ann. Thorac. Surg.* **1998**, *65*, 986–992. [CrossRef]

46. Lichtenberg, A.; Cebotari, S.; Tudorache, I.; Sturz, G.; Winterhalter, M.; Hilfiker, A.; Haverich, A. Flow-dependent re-endothelialization of tissue-engineered heart valves. *J. Heart Valve Dis.* **2006**, *15*, 287–293. [PubMed]

47. Hussein, K.H.; Park, K.M.; Kang, K.S.; Woo, H.M. Heparin-gelatin mixture improves vascular reconstruction efficiency and hepatic function in bioengineered livers. *Acta Biomater.* **2016**, *38*, 82–93. [CrossRef] [PubMed]

48. Badylak, S.F.; Taylor, D.; Uygun, K. Whole-organ tissue engineering: Decellularization and recellularization of three-dimensional matrix scaffolds. *Annu. Rev. Biomed. Eng.* **2011**, *13*, 27–53. [CrossRef] [PubMed]

49. Nietzer, S.; Baur, F.; Sieber, S.; Hansmann, J.; Schwarz, T.; Stoffer, C.; Hafner, H.; Gasser, M.; Waaga-Gasser, A.M.; Walles, H.; et al. Mimicking metastases including tumor stroma: A new technique to generate a three-dimensional colorectal cancer model based on a biological decellularized intestinal scaffold. *Tissue Eng. Part C Methods* **2016**, *22*, 621–635. [CrossRef] [PubMed]
50. Aron, J.; Agarwal, B.; Davenport, A. Extracorporeal support for patients with acute and acute on chronic liver failure. *Expert. Rev. Med. Devices* **2016**, *13*, 367–380. [CrossRef] [PubMed]
51. Selden, C.; Bundy, J.; Erro, E.; Puschmann, E.; Miller, M.; Kahn, D.; Hodgson, H.; Fuller, B.; Gonzalez-Molina, J.; Le Lay, A.; et al. A clinical-scale bioartificial liver, developed for gmp, improved clinical parameters of liver function in porcine liver failure. *Sci. Rep.* **2017**, *7*, 14518. [CrossRef] [PubMed]
52. Erro, E.; Bundy, J.; Massie, I.; Chalmers, S.A.; Gautier, A.; Gerontas, S.; Hoare, M.; Sharratt, P.; Choudhury, S.; Lubowiecki, M.; et al. Bioengineering the liver: Scale-up and cool chain delivery of the liver cell biomass for clinical targeting in a bioartificial liver support system. *Biores Open Access* **2013**, *2*, 1–11. [CrossRef] [PubMed]
53. Selden, C.; Spearman, C.W.; Kahn, D.; Miller, M.; Figaji, A.; Erro, E.; Bundy, J.; Massie, I.; Chalmers, S.A.; Arendse, H.; et al. Evaluation of encapsulated liver cell spheroids in a fluidised-bed bioartificial liver for treatment of ischaemic acute liver failure in pigs in a translational setting. *PLoS ONE* **2013**, *8*, e82312. [CrossRef] [PubMed]
54. Silva, J.M.; Bundy, J.E.E.; Chalmers, S.; Mukhopadhyay, T.; Fuller, B.; Selden, C. *Development of a Small Scale Fluidised Bed Bioreactor for Encapsulated Cell Systems to Be Tested Simultaneously, Thereby Speeding up the r&d Process*; eCM Meeting Abstracts 2016, Collection 5; TCES: London, UK, 2016; p. 73.

bioengineering

MDPI

Article

A Microfluidic System for the Investigation of Tumor Cell Extravasation

Claudia Kühlbach [1,2], Sabrina da Luz [3], Frank Baganz [2], Volker C. Hass [2,4,*] and Margareta M. Mueller [1,*]

[1] Department of Mechanical und Medical Engineering, Hochschule Furtwangen University, Villingen-Schwenningen 78054, Germany; kuec@hs-furtwangen.de
[2] Department of Biochemical Engineering, University College London, London WC1E 6BT, UK; f.baganz@ucl.ac.uk
[3] Hahn-Schickard, Villingen-Schwenningen 78054, Germany, sabrina_daluz@yahoo.de
[4] HFU Hochschule Furtwangen, Department Medical and Life Science, Villingen-Schwenningen 78054, Germany
*   Correspondence: volker.hass@hs-furtwangen.de (V.C.H.); muem@hs-furtwangen.de (M.M.M.); Tel.: +49-7720-3074231 (M.M.M.)

Received: 16 March 2018; Accepted: 21 May 2018; Published: 23 May 2018

Abstract: Metastatic dissemination of cancer cells is a very complex process. It includes the intravasation of cells into the metastatic pathways, their passive distribution within the blood or lymph flow, and their extravasation into the surrounding tissue. Crucial steps during extravasation are the adhesion of the tumor cells to the endothelium and their transendothelial migration. However, the molecular mechanisms that are underlying this process are still not fully understood. Novel three dimensional (3D) models for research on the metastatic cascade include the use of microfluidic devices. Different from two dimensional (2D) models, these devices take cell–cell, structural, and mechanical interactions into account. Here we introduce a new microfluidic device in order to study tumor extravasation. The device consists of three different parts, containing two microfluidic channels and a porous membrane sandwiched in between them. A smaller channel together with the membrane represents the vessel equivalent and is seeded separately with primary endothelial cells (EC) that are isolated from the lung artery. The second channel acts as reservoir to collect the migrated tumor cells. In contrast to many other systems, this device does not need an additional coating to allow EC growth, as the primary EC that is used produces their own basement membrane. VE-Cadherin, an endothelial adherence junction protein, was expressed in regular localization, which indicates a tight barrier function and cell–cell connections of the endothelium. The EC in the device showed in vivo-like behavior under flow conditions. The GFP-transfected tumor cells that were introduced were of epithelial or mesenchymal origin and could be observed by live cell imaging, which indicates tightly adherent tumor cells to the endothelial lining under different flow conditions. These results suggest that the new device can be used for research on molecular requirements, conditions, and mechanism of extravasation and its inhibition.

Keywords: microfluidic device; HPAEC; tumor cell extravasation

1. Introduction

One of the characteristics of malignant cancer is that it can form metastasis in distant organs by tumor cell invasion and the destruction of surrounding tissue [1].

This process is characterized by three indispensable, very complex actions, namely: (i) the dedifferentiation of tumor cells allowing their migration into the metastatic pathways, that is, the circulation [2–6]; (ii) their passive distribution into distant organ systems; and (iii) the transendothelial migration into the surrounding tissue to expand to secondary metastatic tumors [2–6]. The mechanism of extravasation is not yet fully understood, but is thought to resemble the recruitment of leukocytes during an inflammatory response. Critical steps in both processes are the rolling of tumor cells on the inner vessel lining, the tight adhesion to the endothelial cells, and the transendothelial migration [7,8].

Classical cell culture models, while easy to use, do not incorporate the important aspect of cell- and matrix-interactions in a three dimensional (3D) tissue context [9–11]. The 3D cell culture models, which incorporate cell–cell and cell–matrix interactions, and organotypic structures, which more closely resemble the in vivo situation, address this problem [9–11]. A novel approach for 3D cell culture models is the adoption of microfluidic systems, which allow highly reproducible experiments in small volumes of liquids that can be easily controlled [12–14].

1.1. Cancer Metastasis

During the process of metastasis, the intravasation initiates with the increased motility of primary tumor cells that migrate from the primary tumor site to the blood or lymphatic circulatory system [15,16]. When tumor cells reach the vessel, they intravasate a process that requires an active translocation of tumor cells through the barrier of the extracellular matrix and the endothelial lining [15,16]. In the vessel system, the tumor cells are distributed passively, until they reach the metastatic site in the distant organ system, where they extravasate again. This process requires their interaction with surface receptors of the endothelium, which results in a signal transduction that initiates the extravasation process into the surrounding tissue where the tumor cells then create secondary tumors [3,7,15–17]. Only about 1% of the migrating tumor cells establish a distant metastasis [3,7,17]. It is assumed that this process is regulated by the activation and deactivation of several specific genes, including the so called metastasis-suppressor genes, that regulate the development of metastasis but do not influence the tumor growth at the primary site [16,18]. A detailed analysis of the extravasation process reveals three distinct steps, namely: (i) the rolling of cancer cells on the endothelium that activates the endothelial cells, (ii) their tight adhesion to the vessel wall, and (iii) the transmigration through the endothelial monolayer [7,8].

Two different models describe the mechanisms that regulate the adhesion to the vessel wall and extravasation. The 'seed and soil' hypothesis, proposed by Stephen Paget in 1889 [19], claims that the homing of metastatic cells (i.e., seed) requires the interaction with the microenvironment of their target organ (i.e., soil) [15]. Another hypothesis claims that the extravasation entrapment of circulating tumor cells in small capillaries is sufficient [17].

For both models, intimate contact between the tumor cells and endothelial cells is essential to allow adhesion to the vessel wall and subsequent transendothelial migration (TEM). While some aspects of tumor cell extravasation resemble the leukocyte TEM during inflammation, the exact mechanism of contact, adhesion, and TEM of tumor cells are not yet fully understood [7,8]. It becomes abundantly clear that the chemokines and their receptors play a crucial role in every single step of the metastatic cascade [8,16,20], an aspect that can easily be studied in microfluidic devices by the selective addition and blockade of these pathways.

1.2. Microfluidic Devices

Two dimensional (2D) tissue culture models, even when coated with extracellular matrix proteins, are of limited use in mimicking the in vivo conditions [21–23], as they lack any structural or mechanical parameters. In contrast, 3D models that incorporate cells into an extracellular protein matrix, allow the interaction between the tumor cells and their microenvironment [22,23]. One of the most popular static 3D approaches is the Boyden Chamber Assay, where two medium-filled compartments are separated by a porous (sometimes protein coated) membrane that allows the cells to migrate from the upper compartment, through the membrane's pores, into the lower compartment. The number of cells and the time that is needed to reach the lower compartment can be correlated to the malignancy of the cancer cells [24].

However, this model is still not suitable for the investigation of the dynamic effects of cell–cell interactions, spatial organization, and cell migration [22,23], and cannot provide any insights in the complexity of the multistep process of cancer metastasis [21]. To better understand this stepwise process, more sophisticated models are much in need. Microfluidic devices combine the advantage of a 3D-model with dynamic flow conditions, as found in vivo, while allowing standardized, highly reproducible experimental conditions that can provide a basis for high throughput screenings [12–14]. As such, they are well suited for research on the metastatic cascade [25], including the interaction of tumor cells with the endothelial cells of the vessel wall and the influence of forces within the blood stream [21]. In this context, several studies on TEM, using microfluidic models, were published within the last ten years, which suggest an increase in TEM by the application of flow to the tumor cells [26].

One frequently used approach that creates a microvascular network within the microfluidic channel, is taking advantage of the capability of the endothelial cells to create self-assembled tubule-like structures [27–29]. These types of microfluidic models often embed the vascular cells in a matrix of extracellular proteins, like Collagen I, laminin, or fibrinogen [17,30–32]. While the tubule-like structures resemble the capillaries in vivo, they can usually not be subjected to flow and can thus only allow the study of the tumor cell extravasation under static conditions [17,29].

However, there are approaches that allow the application of varying flow rates in an endothelial cell lined vascular equivalent. One possibility is to introduce the endothelial cells as a monolayer into a microfluidic channel, which is often coated with matrix protein-like poly-D-lysine [17,31] or matrigel [33,34], for better adhesion properties. Independent of the model geometry or of the type of endothelial cells that are used, most of these devices are made of Polydimethylsiloxane (PDMS), which is frequently activated by plasma-treatment and is bonded to glass [17,31,33,34]. For example, Zervantonakis e al., 2012, describe a model consisting of two parallel channels that are separated by a hydrogel matrix that contains human endothelial cells from an umbilical cord vein (HUVEC) and fibrosarcoma and breast cancer cells. In this model, the tumor cells migrate towards the endothelial cells and intravasate without adding any kind of flow [35]. A similar model was proposed by Haessler et al., 2012. Here, two adjoining channels were coated with poly-D-lysine and were subsequently filled with hydrogels of bovine collagen type I, either with or without embedded breast cancer cells. Hydrogels of different permeability were used to control the interstitial flow [26], which was shown to influence the migration behavior and migration speed of the breast cancer cells [36].

Another approach to study the adhesion of the tumor cells to an endothelial monolayer was published by Song e al., 2009. The device consists of two PDMS layers with a porous membrane sandwiched in between. The membrane with a pore size of 400 nm prevents the transmigration of the tumor cells, but allows for the diffusion of soluble factors. The upper channel was seeded with human dermal microvascular endothelial cells (HDMVEC), from foreskin and human breast cancer cells, which were introduced into this channel via an inlet. Chemokines could be added to the lower channel under different flow conditions. The expression of CXCL 12 and its corresponding receptor CXCR 4 by the tumor cells was shown to promote tumor metastasis, potentially by a CXCL 12 induced upregulation and activation of adhesion molecules in endothelial cells, which supports the interactions with the circulating tumor cells [36]. Accordingly, in this study, tumor cells preferentially adhere to endothelial cells treated with the chemokine [36].

Using a similar device made from PDMS, Zhang e al., 2012 studied the transmigration of the tumor cells through an endothelial layer, into a second channel that was coated with basement membrane proteins under static conditions [37]. In this microfluidic device, small aggregates of a salivary gland adenoid cystic carcinoma cell line have required the addition of CXCL12 chemokine as a 'homing factor' for successful transmigration [37].

Jeon e al., 2013 presented a similar model with two channels separated by a gel region containing type I collagen. One channel is coated with poly-D-lysine and seeded with HDMVEC and the second channel serves as medium channel. The extravasation of breast cancer cells from the first channel into the gel region of the second one could be observed after one day under static conditions [17].

While these studies were mostly done under static conditions, there are a have been reports describing the addition of flow to microfluidic systems with endothelial cells. In the study by Shin e al., 2011, every second day, 10 mL of buffer was infused at a flow rate of 88 µL/min to the channel [33], whereas Riahi e al., 2014 added tumor cells, at a flow rate of 50 µL/h, to the HUVEC endothelial cell lined suspension flow channel. After adding the tumor cells, the flow rate was lowered to 1 µL/h to allow the adhesion to the endothelial cells. In this study, the tumor cells transmigrated from the suspension flow channel through the endothelial monolayer into a channel containing chemokines in matrigel [34].

Just recently, Cui et al. published a microfluidic device consisting of two independent flow channels with a porous membrane that was sandwiched in between with several cell collection chambers underneath an endothelial monolayer. The membrane, with pore sizes ranging from 10 to 26 µm, was coated with matrix proteins for better adhesion properties and was seeded with primary endothelial cells from the foreskin (HDMVEC). The tumor cells—a human breast cancer cell line—were injected to the overhead chamber and a flow of 20 µL/min of culture medium was added. CXCL12 chemokine was added to the channel underneath the endothelial cell monolayer, which acted as chemo attractant to the tumor cells. After an incubation time of 15 h, the migrated tumor cells were collected and counted. While the system principally seems well suited to study the tumor cell extravasation, the authors describe the confluency of the endothelial cells lining the channel as a major problem. Only the single subareas, where the endothelial cells showed a total confluency on top of the membrane could be used for the analysis of the transmigration of tumor cells [38].

1.3. Aim of the Study

In this study we introduce a new microfluidic device for the analysis of the different processes during the extravasation of tumor cells from blood vessels, and the interaction of the tumor cells with the endothelial lining of blood vessels under dynamic conditions.

Several published microfluidic approaches to address this problem use endothelial cells that have been embedded in hydrogel and take advantage of their ability to self-assemble into tubule-like structures [27–29]. A disadvantage of these systems is the multilayered structure that frequently occurs when the endothelial cells are embedded into a hydrogel (Figure 1A). This results in a non-physiological barrier that the extravasating tumor cells would have to pass through. This does not correspond to the in vivo situation, where the blood or lymphatic vessels contain only a monolayer of endothelial cells [39]. Other microfluidic devices working with endothelial monolayers seed the endothelial cells on top of an extracellular matrix protein coating, either on a porous membrane or on a hydrogel (Figure 1B). The devices usually consist of closed channels, where the endothelial cells' suspension are injected and incubated, until they adhere to the device material. In these devices the confluency of the endothelial cells within the microfluidic channel are often patchy and hard to control, thereby creating some difficulties, since the whole dimension of the microfluidic channel cannot be used for the extravasation experiments [17,31,36,38].

The proposed microfluidic device aims to improve these issues. The device is intended for the investigation of all of the steps of the extravasation process, including the rolling of the tumor cells on the endothelial cells, tight adhesion to the endothelial lining, and transendothelial migration. The microfluidic device will be seeded in a monolayer with primary endothelial cells from the target organ of the metastatic tumor cells used, for example, lung. The endothelial cell confluency can be easily monitored along the whole length and on all of the sides of the microfluidic channel, in order to achieve optimal cell–cell contacts of the endothelial cells and introduced tumor cells. The perfusion of the tumor cells into the endothelial cell lined vessel equivalent can be done under different flow conditions.

We will demonstrate the establishment of a stable endothelial monolayer in the device, without the addition of any matrix proteins, and report the initial experiments of characterizing the tumor cell adhesion under flow conditions.

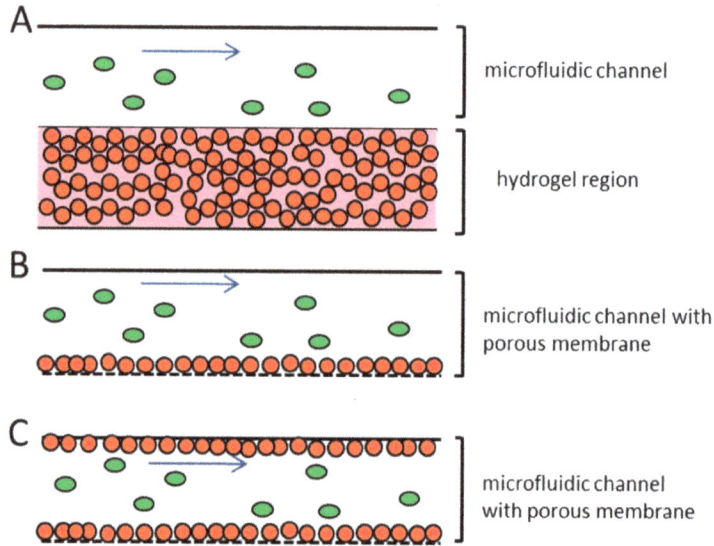

Figure 1. Schematic representation of different microfluidic systems used for the research on the vessel associated steps of the metastatic cascade (**A**) endothelial cells are embedded in hydrogel and injected into the microfluidic channel, where they are arranged in a capillary network. Tumor cells are introduced to the top of the hydrogel, and can migrate into it. No perfusion of the tubule-like structures takes place; (**B**) Endothelial cells are seeded in a monolayer on top of a porous membrane. The confluency of the endothelial cells on the membrane can often not be guaranteed. The tumor cells are introduced to the system under flow conditions; (**C**) Proposed microfluidic device. Endothelial cells are seeded to the total dimension of the microfluidic channel and the adjacent porous membrane, representing the vessel equivalent. The confluency of the endothelial cells is confirmed before the experiment. Tumor cells are introduced to the device under different flow conditions. Red dots—endothelial cells; green ovals—tumor cells; blue arrow—direction of flow.

2. Materials and Methods

2.1. Cells, Cell Lines, and Cell Culture

Human primary pulmonary arterial endothelial cells (HPAEC; Figure A1), green fluorescent protein (GFP) expressing lung carcinoma cells H838 (H838GFP), and SK-Mel 28 malign melanoma cells (SK-Mel 28GFP) were used for this study. The HPAEC were isolated from the pulmonary artery tissue (Pelobiotech, Planegg, Germany). The HPAEC were cultured in T75 culture flasks, with Microvascular Endothelial Cell Medium (Pelobiotech, Germany) and were subcultivated when there was >80% confluent. In order to loosen the cell–cell connections. 0.05% Ethylendiamintetraacetate (EDTA) was used and 0.1% Trypsin/EDTA mix (1:1) was to singularize the cells. The seeding concentration was 5–10 × 10^3 cells per cm^2 growth area. The tumor cells that were used for this study were H838 non-small cell lung cancer cells and SK-Mel 28 melanoma cells. H838 as well as SK-Mel 28 tumor cells were stably transfected with the plasmid pTracerTM-CMV2 vector (Life Technologies, Darmstadt,

Germany) in order to express the GFP and the clones that were characterized to exhibit similar characteristics (proliferation, migration, Boyden Chamber Invasion, 3D organization) to the parental cell line were used for further experiments. The GFP transfected lung carcinoma cell line H838GFP and the GFP transfected melanoma cell line SK-Mel 28GFP cells were cultured in 10 cm culture plates with Dulbecco's Modified Eagle Medium (DMEM) culture medium, which contained 10% fetal calve serum (FCS), 1% penicillin/streptomycin, and 200 µg/mL zeocin as selective antibiotics, at standard conditions (37 °C, 5% CO_2). H838GFP cells were subcultivated by incubation with 10 mL of 0.05% EDTA for 5 min, which was followed by singularizing the cells using 3 mL of 0.025% Trypsin/EDTA mix (1:1) for 3 min. The SK-Mel 28GFP cells were also incubated with 0.05% EDTA for 3 min and singularized, using 0.01% Trypsin/EDTA mix (1:1) for 1 min. The reaction was stopped with a 7 mL culture medium. The seeding concentration of the tumor cells was 3×10^6 cells in a 10 cm culture plate, containing 10 mL of medium.

2.2. Fabrication and Testing of the Microfluidic Device

2.2.1. Identification of Best Suitable Materials

For use in the microfluidic device, different plastic materials were tested for biocompatibility, optical transparency and quality, mechanical processing possibilities, commercial availability, and the potential for steam sterilization. For the experiments, the materials were autoclaved before seeding with HPAEC. Before the seeding process, some of the materials were coated with FCS, collagen, or laminin, so as to improve the cell adherence. HPAEC cells were seeded in a concentration of 7000 to 10,000 cells per cm^2 and were cultivated under standard conditions for a period of at least four days. The results were evaluated photographically and the semiquantitative values (0%, 50% and 100%) for the confluency rates that were achieved after three to four days were allocated.

2.2.2. Device Fabrication

The microfluidic device consisted of three parts (see Figure 2), namely, an upper channel which represented the vessel equivalent, a lower channel to collect the extravasated tumor cells, and a porous membrane (it4ip SA, Louvain-la-Neuve, Belgium), which separated both of the channels. The membrane was made from Polyethylene terephthalate (PET), had a thickness of 19 µm, a pore size of 5 µm, and a pore density of 6×10^4 cm^{-2}. The slides that contained the channels were fabricated by replicate molding of PDMS (Dow Corning, Midland, MI, USA). The mold was manufactured by the Institute of Micro- and Information-Technology of the Hahn–Schickard–Gesellschaft e.V., Germany. After the curing process, the device was oxygen plasma treated and autoclaved. The upper channel had a dimension of 500 µm × 100 µm × 5.9 cm. The endothelial cells were seeded separately on the single parts, which were assembled just before the experiment. The three parts were tightly clamped in an aluminum frame, which made additional sealing (e.g., by bonding of the device) unnecessary. A low pressure syringe pump (Cetoni, Korbußen, Germany) was connected to the upper channel inlet, which allowed continuous or pulsatile flow rates for medium or tumor cell suspension, which ranged from 0.4 to 1.2 µL/s (flow velocities ranging from 8 mm/s to 24 mm/s).

2.2.3. Cell Seeding

The HPAEC were singularized, as described above, and seeded with 1×10^6 cells to the upper channel and the top of the porous membrane, and were incubated for 24 h at standard conditions.

2.3. Tumor Cell Suspension for Perfusion Experiments

For the tumor cell suspension that was perfused into the microfluidic channel, H838GFP or SK-Mel 28GFP cells were subcultured in the DMEM culture medium, which contained 10% FCS and 1% penicillin/streptomycin. The tumor cell concentration for the perfusion of the microfluidic device amounted to 0.5×10^6 cells per mL.

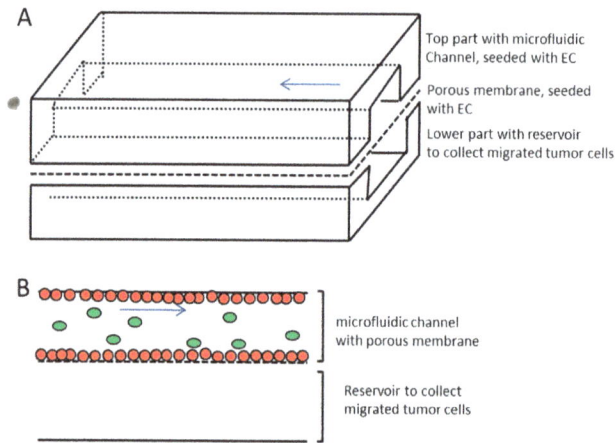

Figure 2. Schematic representation of the microfluidic device. (**A**) Upper and lower channel made of Polydimethylsiloxane (PDMS) with porous membrane (pore size 5 μm) sandwiched in between. The upper channel with membrane represents the vessel equivalent (channel dimensions 500 μm × 100 μm × 5.9 cm, width × height × length, respectively) and is seeded with endothelial cells (EC), the lower channel represents reservoir for transmigrating tumor cells (lower channel dimensions 1 mm × 1 mm × 5.9 cm, width x height × length, respectively). (**B**) Side view sketch of microfluidic channels and membrane separating the two channels. Red dots—endothelial cells growing on all sides of the upper channel and on the membrane; green ovals—tumor cells introduced to vessel equivalent; blue arrow—direction of flow.

2.4. Flow Experiments

The assembled microfluidic system was connected to the low pressure syringe pump. The maximum flow velocity for the medium flow or influx of the tumor cell suspension ranged from 8, 12, 16, to 24 mm/s, either with a continuous or pulsation frequency of 60/min. The pulse duration was 500 ms, which was followed by a stop of flow for 500 ms. The duration of the experiment ranged from 6 to 72 h, as stated in the results section.

After finishing the experiment, the microfluidic device was disassembled for further analysis, such as immune fluorescence staining or counting of adherent tumor cells in different areas over the whole length of the microfluidic channel, which omitted the inlet and the outlet. For the disassembly, the clamp that held the aluminum frame was taken off and the microfluidic device and the three single parts were taken apart.

2.5. Immunofluorescence

Before the staining, the probes were incubated in 4% paraformaldeyde (PFA, in 1× phosphate buffered saline (PBS)) for 20 min, which was followed by the treatment with 70% and 100% ethanol for 5 min each. The blocking of the probes was executed with 18% bovine serum albumin (BSA) for 1 h. The primary antibodies were incubated overnight at 4 °C. For the negative control, just 18% BSA without the primary antibody was added to the probe. The negative control was stained on the same probe, either slides, or the material of the microfluidic device. The experimental conditions were the same for the negative and the positive group. The next day, three washing steps with 1× PBS for 5 min, to remove the unbound primary antibodies, were carried out. The secondary antibodies were added and incubated for 1 h at room temperature. After three more washing steps, the nuclei were stained with Hoechst dye (Sigma-Aldrich, Albuch, Germany) and were diluted 1:1000 with 1× PBS for 5 min. This was followed by three more washing steps with 1× PBS, which was followed by embedding the probes with Dako Fluorescent Mounting Medium (Agilent, Waldbronn, Germany). The staining was microscopically evaluated. Primary antibodies used were Anti-VE-Cadherin in a concentration of 1:200 in 18% BSA (from rabbit, Sigma-Aldrich, Germany) for the detection of adherence junctions, which were the cell–cell connections in the endothelial cells [40] and Anti-Collagen IV in a concentration of 1:50 in 18% BSA (from rabbit, Progen, Heidelberg, Germany) for the detection of the extracellular matrix protein collagen IV. This was the main component of the human basal membrane [41,42]. The secondary antibody was used in a concentration of 1:500 in 18% BSA (Cy 3 donkey, anti-rabbit, Dianova, Hamburg, Germany).

2.6. Data Analysis

The number of adherent tumor cells to the endothelial lining was determined by counting the adherent tumor cells at the top of the channel and on the membrane on a length of 2 mm at five different positions, each. The mean values were taken from the 10 single counts per microfluidic device and the results were extrapolated to mean values/mm^2. The standard deviation was taken over the different devices that were used in the experiment. Either two or three replicate experiments were performed, as stated in the results section.

3. Results

3.1. Design and Fabrication of the Microfluidic Device

Prior to establishing the microfluidic system as a model for micro capillary vessels, experiments were conducted in order to determine the most suitable materials, with respect to biocompatibility. The material should not have affected the proliferation and maintenance of the endothelial monolayer, but also should not have been degraded by the cells. Different materials were tested and the most suitable was chosen, based on the confluency levels after a cultivation time of 3–4 days.

The material selection was also influenced by the ease of the device assembly, since efficient seeding and assessment of cell confluency in the microfluidic system was simplified by fabricating a device that was made up of three different parts (see Figure 1).

The most suitable materials were cyclo-olefin polymer (COP) and cyclo-olefin co-polymer (COC), with 100% confluency of endothelial cells and PDMS with a cell confluency of 70% after three days. COP and COC are quite expensive materials and because of their structure, they would need to be thick in the cross-section, in order to be able to cut the microfluidic channels into the surface. This increased material thickness would, however, restrain the transparency of the device and thus restrict the live microscopic observation in the channels. Additionally, a device that was made from these materials would need to have additional seals between the different parts so as to avoid leakage after assembly. PDMS was chosen because it was less expensive, easier to manufacture, and there was no need for additional sealing of the parts. Furthermore, after the surface activation of the PDMS oxygen plasma treatment, a confluency rate of 100% was achieved in a time frame that was comparable to that using COC or COP, making PDMS the most suitable material for the microfluidic system, with regards to all of the design requirements.

The microfluidic device was comprised of three different parts. Two parts were made from PDMS, which contained the microfluidic channels. The upper channel, which represented the vessel equivalent, had a size of 500 μm × 100 μm × 5.9 cm (width × height × length). This channel was bordered on the lower side by a porous membrane with a pore size of 5 μm. The channel surface and the neighboring membrane were seeded with endothelial cells. The lower channel functioned as a cell trap and had a size of 1 mm × 1 mm × 5.9 cm (width × height × length). Immediately before the experiment, the three parts were mounted into an aluminum frame and infused with endothelial cell medium or the tumor cell suspension. As a result of the PDMS material properties, the additional bonding of the different parts was not necessary. The seeding of the single parts of the microfluidic device in an open setting before assembly was chosen for better control of homogeneous population and confluency of the endothelial cells. A tight confluent monolayer of endothelial cells, which was achieved after 24 h of cultivation in an open setting, were required before the assembly and the start of any experiment, in both the microfluidic channel, as well as on the adjacent membrane. This was essential to assure a tight barrier function of the endothelium along the entire dimension of the microfluidic channel for the extravasation experiments.

Coating of the Microfluidic Channel and Membrane

In the circulatory system, the inner vessel wall was arranged as monolayer endothelial lining, which rested on a basement membrane was composed of different extracellular matrix proteins, mainly collagen IV [43,44]. To better represent the vessel wall in the microfluidic device, the vessel equivalent was coated with different extracellular matrix proteins, like collagen I and laminin. However, within 24 h, the coating was disintegrated by the HPAEC endothelial cells that were used in this study (Figure A1). An anti-collagen IV immune fluorescence staining of the endothelial monolayer cultures in the open device revealed that the endothelial cells secreted collagen IV, and thus established their own basement membrane to rest on, which eliminated the need for coating with a supporting matrix (Figure 3).

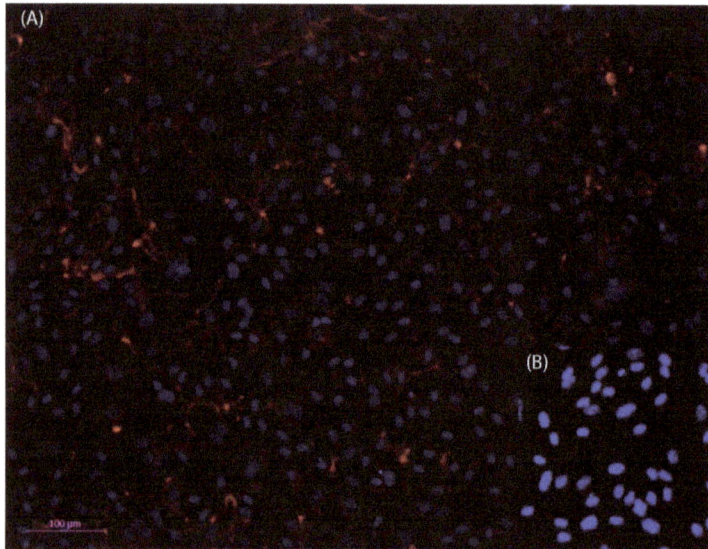

Figure 3. Fluorescent microscopic photos of anti-Collagen IV, Cy3 (red) immunofluorescence staining of human primary pulmonary arterial endothelial cells (HPAEC) (**A**); nuclei stained with Hoechst dye (blue). Collagen IV can be detected both in the cytosol of the cells as well as outside of the cells, which indicates that the HPAEC cells have formed an adequate basement membrane (**B**). The negative control shows that the method has been used appropriately. Dilution of antibodies: 1st antibody anti collagen IV 1:50, and 2nd antibody Cy3 anti rabbit 1:500). Scale bar 100 μm.

3.2. Stability of the Endothelial Cell Monolayer within the Microfluidic Device

3.2.1. Treatment of the Endothelial Cells with Flow

The functionality of the endothelial cell lining in the microfluidic channel was confirmed by perfusing the vessel equivalent channel with endothelial cell medium for up to 72 h, which used flow rates of 0.8 or 1.2 μL/s. The perfusion was either continuous or with a pulsation rate of 60/min. Figure 4 shows the endothelial lining within the microfluidic vessel equivalent, before and after perfusion.

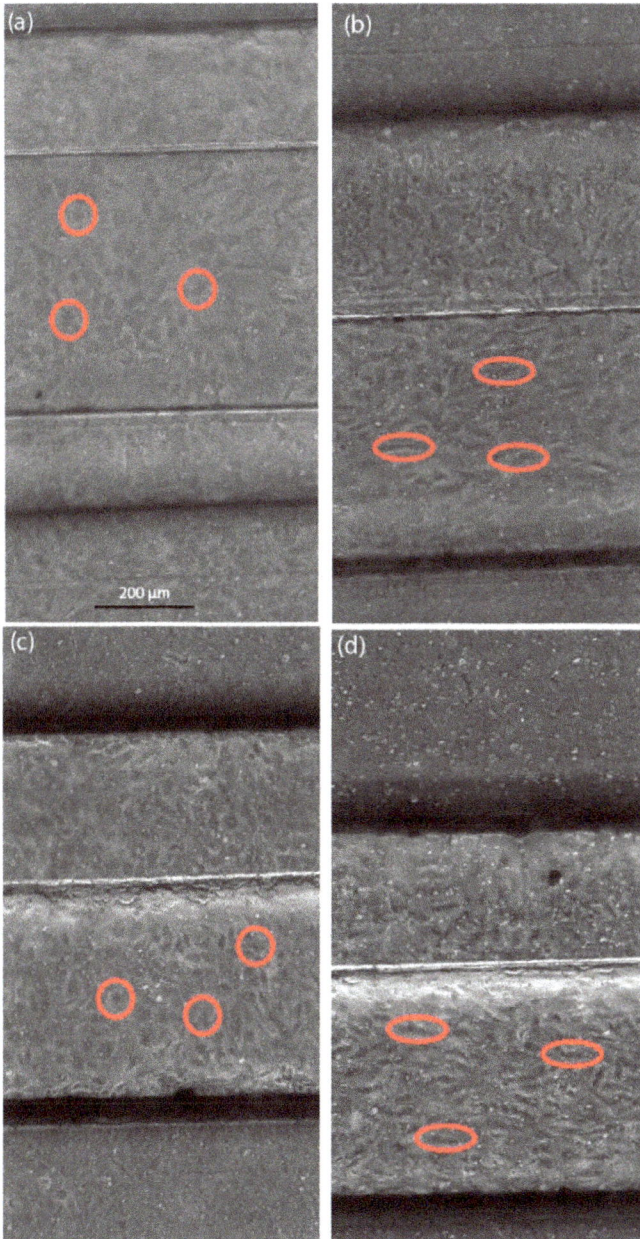

Figure 4. Live cell images of endothelial cells in the microfluidic device, perfusion flow velocity 24 mm/s. (**a**) Image of endothelial cells at the start of the continuous flow experiment; (**b**) Endothelial cells after continuous flow over 48 h; (**c**) Image of endothelial cells at the start of the pulse flow experiment; (**d**) Endothelial cells after the pulse flow perfusion at a rate of 60/min over 72 h. Before the flow is introduced, endothelial cells have a roundish shape (red circles). Endothelial cells show morphological changes, like elongation and orientation, into the direction of flow (red ovals), which might be influenced by the shear stress of the continuous or pulsating flow.

Endothelial cells showed an elongated shape and an organization of the cell distribution in the flow direction. These morphological changes took place during the continuous as well as pulsation flow of 60/min, independent of the flow rate that was used. However, the changes took a longer time to become visible at lower flow rates. For the following experiments, the endothelial cell lining was treated with a continuous flow before the introduction of any tumor cells into the system.

3.2.2. Immunofluorescence of VE-Cadherin

The VE-Cadherin was a transmembrane protein that was involved in the mechanical stability of endothelial tissue and that also played an important role in the signal transduction pathways [45–47]. VE-Cadherin was an adherence junction protein and was expressed all over the cell surface area. On the inside of the cell membrane, VE-cadherin was connected to the actin cytoskeleton via catenin [45]. The membrane associated localization of the VE-Cadherin by immunostaining was considered as an adequate marker for the regular assembly of endothelial cell junctions. In the system that was introduced here, a VE-Cadherin immuno fluorescence staining was performed, which revealed a protein localization in the cell membrane, as an antibody for VE-Catherin was used, which indicated that cell–cell contacts were regularly expressed and integrated into the endothelial cell membrane. Figure 5 shows images of the endothelial lining in the microfluidic device after an anti-VE-Cadherin immunofluorescence staining.

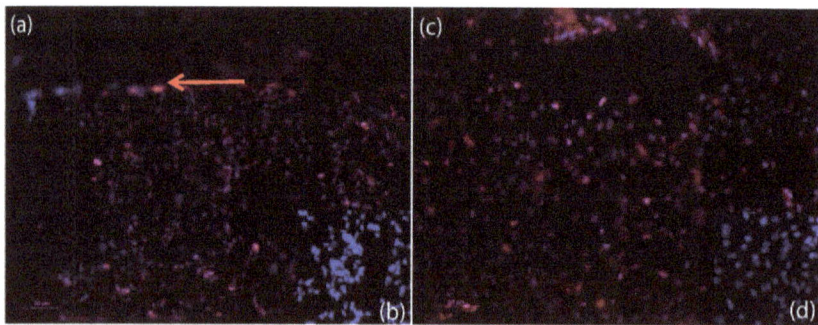

Figure 5. Fluorescent microscopic photos of anti-VE-Cadherin, Cy3 (red) immunofluorescence staining of HPAEC endothelial cells in the microfluidic device after finishing an experiment (nuclei stained with Hoechst) (blue). (**a**) Anti-VE-Cadherin Staining of the HPAEC endothelial cells within the microfluidic channel after an experiment. Red arrow shows edge of channel. VE-Cadherin can be detected on the surface of all endothelial cells and shows adherence junction expression; (**c**) Anti-VE-Cadherin staining of the HPAEC endothelial cells on top of the porous membrane of the microfluidic device after an experiment. VE-Cadherin can be detected on the surface of all of the endothelial cells and shows adherence junction expression. (**b,d**) Negative control for (**a,c**) show that method was conducted appropriately. Dilution of antibodies: 1st antibody anti VE-Cadherin 1:200, and 2nd antibody Cy3 anti rabbit 1:500). Scale bar: 50 μm.

3.3. Introduction and Adhesion of the Tumor Cells on the Endothelial Lining

The endothelial monolayer was incubated and perfused with medium for 30 min before the tumor cells were introduced to the system so as to ensure tight cell–cell contacts because of the exposure of flow in the microfluidic channel. The tumor cell suspension was infused for the duration of 6 h. After the end of the tumor cells circulation, the microfluidic system was disassembled and was fixed with 4% PFA for further analysis. The membranes and the microfluidic channels were microscopically analyzed and the adherent tumor cells that were easily identified because of the expression of GFP were counted (Figure 6, red arrows).

Immediately before the experiment, the endothelial cells' lined channel was perfused with a medium at a continuous flow velocity of 24 mm/s for 30 min. The tumor cells were introduced into the system with different flow rates, which ranged from 0.4 to 1.2 µL/s and corresponded to a flow velocity from 8 to 24 mm/s, respectively. The different experiments included a continuous and pulsatile flow. The tumor cells that adhered to the endothelial monolayer within the vessel equivalent were counted and the numbers were compared (Figure 7).

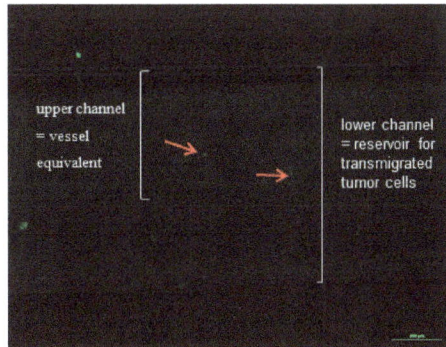

Figure 6. Example of an image of the microfluidic channel with endothelial lining (HPAEC) and adherent tumor cells (H838GFP) during an experiment, after a flow time of 60 min at a flow velocity of 24 mm/s. Endothelial cells were incubated with medium for 30 min before tumor cells were introduced into the microfluidic system.

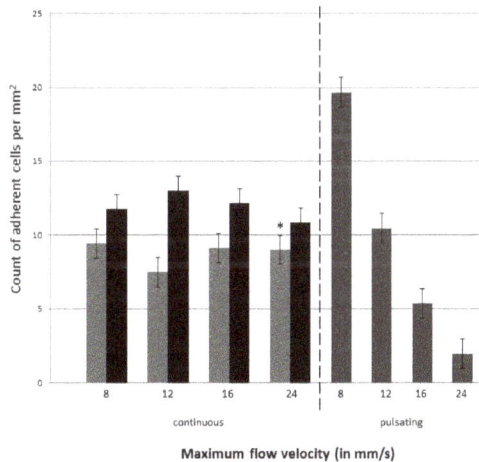

Figure 7. Number of H838GFP and SK-Mel 28GFP tumor cells adherent to the endothelial monolayer within the vessel equivalent. Light grey bars show the count for H838GFP, black bars the count of SK-Mel 298GFP for continuous flow, dark grey bars represent the number of H838GFP cells adherent to the endothelial cells when perfused with pulsatile flow. Four different flow velocities of 8, 12, 16, and 24 mm/s were used in both the continuous and pulsatile flow. Pulsation rate was 60/min. Endothelial cells were incubated with a medium for 30 min before the tumor cells were introduced. Tumor cell suspension was infused for 6 h. The mean value represents the average number of adherent cells in five different areas of 2 mm length in the microfluidic channel (top of the channel and membrane). The standard deviation is taken over the number of microfluidic devices used for the measurements. Dark grey bars and *: three devices were analyzed; light grey and black bars: 2 devices were analyzed.

The data indicated that during the continuous flow, the number of adherent tumor cells in the endothelial cell-lined channel were not influenced by the flow rate, when they were perfused with a continuous flow. In contrast, when tumor cells were introduced into the system with a pulsating flow, more H838GFP cells adhered to the endothelial cells when the flow rate/velocity was lower. When comparing the number of adherent tumor cells at the same flow velocity, during either the continuous or pulsating flow, the number of cells was higher during the pulsating flow up to a flow velocity of 12 mm/s. At higher flow velocities, the number of adherent tumor cells were higher when they were perfused with a continuous flow.

In addition to the H838GFP lung carcinoma cells that were of epithelial origin, the microfluidic device was also perfused with SK-Mel 28GFP, which were cells of a mesenchymal origin, at a concentration of 5×10^5 cells per mL. As for the H838GFP cells, the flow rate did not seem to influence the number of adherent cells to the endothelium under continuous flow. Interestingly, the comparison of the number of adherent H838GFP lung carcinoma and SK-Mel 28GFP melanoma cells to the endothelium, after an infusion time of 6 h, revealed a lower number of adherent lung carcinoma cells than of the adherent melanoma cells, independent of the flow rate that was used (Figure 7).

4. Discussion

For research on the metastatic cascade, it was crucial to develop 3D models of endothelial lined vascular systems, which could be perfused with either medium or tumor cells [14,23,26,31], and allowed an easy manipulation of the flow rates and shear stress in the vascular channel. Using this type of microfluidic system, the transendothelial migration, an essential step in successful metastasis of tumor cells, could be investigated under in vivo-like conditions [48]. To this end, a PDMS-based microfluidic system was introduced in this work. The device was assembled from three different parts, namely, two parallel channels with a porous membrane sandwiched in between. The membrane has a pore size of 5 µm and acted as the area for the transmigration of the tumor cells, while at the same time allowing a confluent population with endothelial cells, which mimicked the vessel wall. In the microfluidic devices that were introduced previously, the area for the transmigration was either defined by the microgaps of different sizes [49]; porous membranes with different pore sizes, ranging from 10 µm to 26 µm [31,36,38]; or the tumor cells and had to extravasate through the endothelial cells that were embedded in a hydrogel matrix [17,31,32,37]. These devices were limited in their comparability to the in vivo situation, since the establishment of a confluent endothelial cell layer on an authentic endothelial cell that was derived from ECM and a regular expression of endothelial cell junction proteins could not be guaranteed.

In the device that was introduced here, the upper channel and the membrane were seeded with an endothelial cell monolayer in an open stage manner, prior to the assembly of the system. The concept of the open staged seeding and culture prior proved to be highly useful for achieving a confluent endothelial cell layer along the complete channel walls, which could easily be observed and controlled over the entire length of the microfluidic channel and the membrane that represented the vessel equivalent. The other microfluidic systems that were published were constructed with closed microfluidic channels, where seeding the of the endothelial cells took place by injecting the cells into the channel [17,30,36–38]. In these devices, the cells were incubated within the microfluidic channel from 10 min to 1 h, so as to ensure the adhesion of the cells to the coating of the microfluidic device [17,36,37]. Frequently, these studies failed to confirm a homogeneous population of the devices or to describe the difficulties in achieving a complete endothelial cell coverage. A control for the confluency of the endothelial monolayer was solely described by Jeon e al., using microscopic observation, before starting the extravasation experiments [17]. Cui et al. even described the technical difficulties so as to assure the confluent cell coverage over the whole membrane area [38]. These problems were resolved by the model that has been described here, since the control of a confluent endothelial monolayer throughout the entire device could be achieved microscopically before mounting the different parts to an experimental frame.

To ensure a high comparability to the in vivo situation, the endothelial cells that were used in this study were primary endothelial cells of the lung. The microfluidic vascular equivalents that were coated with these cells optimally represented the in vivo situation in metastasizing tumors—since the lung is one of the major homing sites for metastatic tumor cells of different origin. In contrast, the devices that were previously described by the other groups used human umbilical cord vein endothelial cells (HUVEC) [17,31,32] or human dermal microvascular endothelial cells (HDMVEC) that were isolated from the foreskin [31,36,38]. The seed and soil hypothesis, which was first published by Paget 1889, stated that the distribution of metastasis was not coincidental, but that different organs were 'predisposed' for the secondary tumor growth [19,50]. The establishment of a secondary tumor was thought to depend on the molecular communication of the tumor cells with the specific organ microenvironment, which included the endothelial cells that supported the survival and growth of the tumor cells [15,51]. Thus, because of the potential heterogeneity of the endothelial cells of the different tissues, the use of the primary endothelial cells, which were isolated from the metastatic site of interest, were to be preferred. Indeed, the studies showed significantly differed protein expression profiles in the HUVEC and endothelial cells of other origin, as well as the different behaviors of the foreskin endothelial cells versus the capillary endothelial cells from other organs [52]. While the HUVEC endothelial cells were frequently used in the model systems, the metastasis through the umbilical cord was a rare event with reliable data still lacking. Similarly, metastasis to the skin was very rare, with half of them being the outcome of the outgrowing tumor mass of the underlying primary tumor [53]. In contrast to these data, the metastasis to the lung was commonly seen in a number of tumors, such as breast, colorectal, kidney, head/neck, testicular and bone carcinomas, sarcomas, melanomas, and thyroid cancer [54,55]. Therefore, the HPAEC endothelial cells were chosen for the population of the microfluidic device in order to allow a good representation of the metastatic environment in vivo.

To mimic the blood vessel structure in vivo, it was important that the endothelial cell monolayer adhered to a basement membrane that was made up of extracellular matrix proteins, such as collagen IV and laminin, in the microfluidic vessel equivalent [56]. As a consequence of the microfluidic systems, which were established for the study of the metastatic cascade and were used for coating of the devices, mostly for better adhesion properties. The extracellular matrix proteins were most frequently used for coating either a collagen I hydrogel, matrigel, or poly-D-lysine [17,25,26,30,31,34–36,38]. While both collagen I and matrigel were not regular components of the vascular basement membrane [43], their use for coating was very common. To adapt the system that was presented here to a basement membrane, like the surface underneath the endothelial cells that coated the vessel equivalent microfluidic channel and porous membrane with ECM proteins, was tested. However, since the coating did not reliably adhere to the material and the endothelial cells disintegrated the extracellular matrix coating within 24 h after seeding, this approach was abandoned. The endothelial cells had been known to secrete a matrix metalloproteases to disintegrate and rebuild the extracellular matrix during angiogenesis and vascular remodeling [44]. The results that were obtained here suggested that the establishment of a confluent monolayer of endothelial cells in the channel was associated with the mechanisms that were seen during the vascular remodeling, which ultimately led to the disintegration of the extracellular matrix coating. In vivo, the endothelial cells were known to secrete components of the basement membrane [31], which mostly contained collagen IV fibers [25,45,46]. To determine whether the endothelial cells in the microfluidic device were able to establish their own basement membrane, an anti-collagen IV immune-fluorescence staining was performed on the HPAEC cells that were seeded in the vessel that was equivalent of the microfluidic device that was used in this work. The staining showed a partial staining for collagen IV in the cytosol of the endothelial cells, but also verified the secretion of collagen IV to the growth surface of the microfluidic system. Thus, it could be assumed that the endothelial cells that were used in this study were capable of establishing their own basement membrane, which made an additional coating of the device with the extracellular matrix proteins unnecessary. Thus, the system provided an authentic ECM for the attachment of the endothelial cells, which should have been better able to represent the in vivo environment. Specifically, collagen I and matrigel, which were frequently used in the microfluidic systems, were not regular components of the vascular basement membrane and could show the batch to batch variability and could therefore be a source of variable experimental results [43,47].

This was further confirmed by a regular expression of the cell junction protein VE-Cadherin at the cell surface of the endothelial cells that lined the microfluidic channel. The regular expression of the cell junction proteins, among them VE-Cadherin, was an essential characteristic of an intact endothelial cell lining in the vasculature [40]. While some publications failed to test the integrity of the endothelial cell monolayer in the system before the tumor cells were introduced [37], the analysis of the VE-Cadherin expression was a well-established method for microfluidic devices that used endothelial cells in order to examine the integrity of the endothelial cell monolayer. The regular staining against the VE-Cadherin all over the cell surface, without any gaps, which suggested a proper expression of cell–cell contacts in the system that was described here, was in agreement with the results from other research groups that characterized the endothelial cell lining of the microfluidic devices [17,34]. As such, the system provided an excellent platform for a dynamic capillary model.

In the new dynamic device that was introduced here, the seeded endothelial cells in the vessel equivalent could be exposed to flow, similarly to the in vivo situation. As a result, the endothelial cells showed morphological changes from a polygonal appearance to a more ellipsoid one, under high flow velocities. Additionally, they exhibited an orientation along the direction of the flow. Previous studies described similar changes upon the exposure of endothelial cells to the flow velocities and showed that the changes in the phenotype happened earlier than the changes in the cell alignment [57,58]. These variations and morphological changes in the appearance of the endothelial cells could have also been observed in the device that was used in this work. When adding a medium flow to the endothelial cell monolayer, within the vessel equivalent for more than 24 h, the changes of the phenotype and alignment occurred later during the pulsatile than during continuous flow, an observation that agreed with the study of Adams and Shaw [59].

In the microfluidic system that was established in this work, the tumor cells that were introduced to the endothelial cell-lined vessel equivalent were transfected with GFP, which enabled the identification of the tumor cells during the system perfusion via the live cell imaging, using a fluorescence microscope. The introduction of tumor cells into the system occurred at a flow rate, ranging from 0.4 to 1.2 µL/s, in either the continuous or pulsatile mode. These flow rates corresponded to the flow velocities of 8 to 24 mm/s. In the human aorta, the mean flow velocity was around 11 cm/s, however this vessel had a diameter of about 3 cm [60]. The flow velocities that were used for the microfluidic device that has been presented here, were much lower, yet the microfluidic channel, which acted as a vessel equivalent, only had the dimensions of 500 µm in width and was 100 µm high. In contrast, a human capillary vessel only has a diameter of 40–100 µm and a flow velocity of around 0.3 mm/s [60]. While the capillary flow velocity was significantly lower than the flow velocity that was used in the experiments that have been reported here, it is noted that the microfluidic channel had much larger dimensions, thus reducing the shear stress in comparison to a capillary at the same flow velocity. Other studies reported flow rates of 88 µL/min in a microfluidic channel with 2 mm width and 75 µm height [33]. This corresponded to a flow velocity of 10 mm/s, which was near the minimum velocity that had been tested with the microfluidic device that was presented here. Thus, the system that has been presented allowed for the application of a considerable shear stress and thus provided the basis for a systematic analysis of the influence of the shear stress and flow velocity on the tumor cell attachment to the endothelial cell lined vessel wall.

During the perfusion of the tumor cells through the microfluidic channel, the rolling as well as tight adhesion of these cells to the endothelial cells could be observed. The rolling process was characterized as loose adhesions that were broken off by the dynamic flow [7,61,62] and were considered a prerequisite for the tight adhesion and potential subsequent transendothelial migration of the tumor cells. While the flow velocity seemed to influence the tumor cell adhesion during the pulsatile flow, at a continuous flow—which was likely to be found in microcapillaries in vivo—the number of adherent tumor cells to the endothelium did not seem to depend on the flow velocity through the microfluidic system. Similar results were described by others [7,63]. In the work of Cui e al., the tumor cells were seeded on top of the endothelial cells and perfused afterwards with a rate of 20 μL/min of medium [39]. The transendothelial migration could be observed within 15 h, without flow [17], or after 24 h when the flow was applied to the microfluidic device [38]. In the experiments that were presented here, the rolling and the tight adhesion could be observed after 6 h. A transendothelial migration could not be observed within this time range, which make the cell trap excessive up to this time. In future tests, the infusion of the tumor cell suspension should have been extended to at least 24 h, so as to monitor the transendothelial migration processes.

5. Conclusions

In this study, a new microfluidic device is introduced in order to improve the issues occurring in the devices that have already published, particularly the control of the total confluency along the microfluidic unit, variabilities of cell behavior because of the extracellular matrix protein coating, and the use of primary endothelial cells from the metastatic target organs, which are more suitable for the research on metastatic processes. The devices that are used for research on tumor metastasis include the systems where the multilayer endothelial cells are embedded into a hydrogel to build a tubule-like vascular network. In other devices, the endothelial cells are seeded in the monolayer on top of an extracellular matrix protein coating along a microfluidic channel and the tumor cells can be introduced under flow conditions. These devices are mostly difficult to control for the confluency of the endothelial lining before starting an experiment.

The microfluidic device that has been introduced here consists of three parts, namely, two microfluidic channels and a porous membrane sandwiched in between. The upper, smaller channel, and the membrane acts as vessel equivalent and is seeded with primary endothelial cells that are isolated from the lung artery. This cell type was chosen since the lung is a favored site for the metastasis for many cancer types. The lower channel acts as reservoir to collect the extravasated tumor cells. The parts for the vessel equivalent can be seeded separately, with the endothelial cells in a concentration that is high enough to ensure a confluent monolayer over the whole length of the microfluidic channel. Confluency is controlled before the assembly of the device and at the start of any experiment. An additional coating of the device is not necessary for the endothelial cells that are used, as they secrete their own matrix within 24 h. The endothelial cell monolayer integrity was investigated using an anti-VE-Cadherin immuno-fluorescence staining and showed tight cell–cell contacts between the single cells of the monolayer. Under the flow conditions, the endothelial cells exhibited in vivo-like behavior, including the elongation of the cells and change of orientation in the direction of the flow. The tumor cells that were used for the study were the cancer cells of both epithelial and mesenchymal origin. The GFP transfected lung carcinoma cells H838 and malignant melanoma cell line SK-Mel 28 were introduced to the device as single cell suspension, under different flow conditions. The maximum flow rate ranged from 0.4 to 1.2 μL/s, using either a continuous flow or pulsatile flow with a rate of 60/min. The results show that the cancer cells adhere tightly to the endothelium under these conditions. In continuous mode, the number of adherent cells does not seem to depend on flow rate. Transendothelial migration could not be observed, as the experiments were terminated 6 h after the tumor cell introduction.

In summary, our results suggest that the device that has been introduced here can be used for the research on tumor cell extravasation and the mechanism of rolling, adhesion, and transendothelial migration of metastatic cells. The studies of chemokines, like CXCL 12 and TNF-α, as homing factors or adhesion inhibitors influencing the extravasation process, can be done by adding them either to the tumor cell suspension or into the reservoir, so as to collect the transmigrated tumor cells.

Author Contributions: C.K. carried out the experiments. C.K. wrote the manuscript with support from F.B., V.C.H. and M.M.M., S.d.L. fabricated the microfluidic devices. M.M.M. and V.C.H. helped supervise the project. M.M.M. conceived the original idea. V.C.H. and M.M.M. supervised the project.

Funding: The project was funded by a BMBF grant (FKZ 031A255A MICROMET) to Hahn-Schickard and M.M.M.

Conflicts of Interest: The authors declare no conflicts of interest.

Appendix A

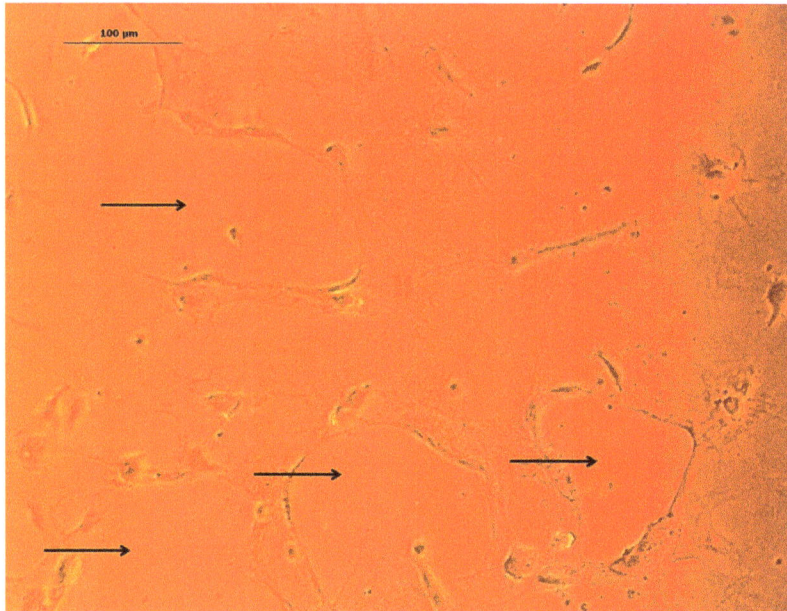

Figure A1. Example image of HPAEC endothelial cells disintegrating collagen coating after incubation time of 24 h. Within the culture, holes within the collagen layer can be observed (black arrows). Scale bar 100 µm.

References

1. Ramaswamy, S.; Ross, K.N.; Lander, E.S.; Golub, T.R. A molecular signature of metastasis in primary solid tumors. *Nat. Genet.* **2002**, *33*, 49–54. [CrossRef] [PubMed]
2. Piris, A.; Mihm, M.C. Mechanisms of metastasis: Seed and soil. In *Cancer Metastasis and the Lymphovascular System: Basis for Rational Therapy*; Leong, S.P.L., Rosen, S.T., Eds.; Springer Science & Business Media: New York, NY, USA, 2007; pp. 119–127. ISBN 978-0-387-69219-7.
3. Li, S.; Sun, Y.; Gao, D. Role of the Nervous System in Cancer Metastasis. *Oncol. Lett.* **2013**, *5*, 1101–1111. [CrossRef] [PubMed]
4. Radinsky, D.C. Epithelial-mesenchymal transition. *J. Cell Sci.* **2005**, *118*, 4325–4326. [CrossRef] [PubMed]
5. Gavert, N.; Ben-Ze'ev, A. Epithelial—Mesenchymal transition and the invasive potential of tumors. *Trends Mol. Med.* **2008**, *14*, 199–209. [CrossRef] [PubMed]
6. Guadamillas, M.C.; Cerezo, A.; Del Pozo, M.A. Overcoming anoikis—Pathways to anchorage independent growth in cancer. *J. Cell Sci.* **2011**, *124*, 3189–3197. [CrossRef] [PubMed]
7. Giavazzi, R.; Foppolo, M.; Dossi, R.; Remuzzi, A. Rolling and Adhesion of Human Tumor Cells on Vascular Endothelium under Physiological Flow Conditions. *J. Clin. Investig.* **1993**, *92*, 3038–3044. [CrossRef] [PubMed]

8. Miles, F.L.; Pruitt, F.L.; van Golen, K.L.; Cooper, C.R. Stepping out of the flow: Capillary extravasation in cancer metastasis. *Clin. Exp. Metastasis* **2008**, *25*, 305–324. [CrossRef] [PubMed]

9. Kim, J.B.; Stein, R.; O'Hare, M.J. Three-dimensional in vitro tissue culture models of breast cancer—A review. *Breast Cancer Res. Treatm.* **2004**, *85*, 281–291. [CrossRef] [PubMed]

10. Kim, J.B. Three-dimensional tissue culture models in cancer biology. *Semin. Cancer Biol.* **2005**, *15*, 365–377. [CrossRef] [PubMed]

11. Ravi, M.; Paramesh, V.; Kaviya, S.R.; Anuradha, E.; Solomon, F.D.P. 3D Cell Culture Systems: Advantages and Applications. *J. Cell. Physiol.* **2015**, *230*, 16–26. [CrossRef] [PubMed]

12. Bruus, H. *Theoretical Microfluidics*; Oxford University Press: Oxford, UK; New York, NY, USA, 2008; pp. 1–7. ISBN 978-0199235094.

13. Dittrich, P.S.; Manz, A. Lab-on-a-chip: Microfluidics in drug discovery. *Nat. Rev. Drug Discov.* **2006**, *5*, 210–218. [CrossRef] [PubMed]

14. Wlodkowic, D.; Cooper, J.M. Tumors on chips: Oncology meets microfluidics. *Curr. Opin. Chem. Biol.* **2010**, *14*, 556–567. [CrossRef] [PubMed]

15. Fidler, I.J. The pathogenesis of cancer metastasis: The 'seed and soil' hypothesis revisited. *Nat. Rev. Cancer* **2003**, *3*, 453–458. [CrossRef] [PubMed]

16. Kakinuma, T.; Hwang, S.T. Chemokines, chemokine receptors, and cancer metastasis. *J. Leukoc. Biol.* **2006**, *79*, 639–651. [CrossRef] [PubMed]

17. Jeon, J.S.; Zervantonakis, I.K.; Chung, S.; Kamm, R.D.; Charest, J.L. In Vitro Model of Tumor Cell Extravasation. *PLoS ONE* **2013**, *8*, e56910. [CrossRef] [PubMed]

18. Fidler, I.J.; Radinsky, R. Genetic Control of Cancer Metastasis. *J. Natl. Cancer Inst.* **1990**, *82*, 166–168. [CrossRef] [PubMed]

19. Paget, S. The distribution of secondary growths in cancer of the breast. *Lancet* **1889**, *133*, 571–573. [CrossRef]

20. Konstantopoulos, K.; Thomas, S.N. Cancer cells in transit: The vascular interactions of tumor cells. *Annu. Rev. Biomed. Eng.* **2009**, *11*, 177–202. [CrossRef] [PubMed]

21. Van Marion, D.M.S.; Domanska, U.M.; Timmer-Bosscha, H.; Walenkamp, A.M.E. Studying cancer metastasis: Existing models, challenges and future perspectives. *Crit. Rev. Oncol. Hematol.* **2016**, *97*, 107–117. [CrossRef] [PubMed]

22. Sung, K.E.; Beebe, D.J. Microfluidic 3D models in cancer. *Adv. Drug Deliv. Rev.* **2014**, *79–80*, 68–78. [CrossRef] [PubMed]

23. Van Duinen, V.; Trietsch, S.J.; Joore, J.; Vulto, P.; Hankemeier, T. Microfluidic 3D cell culture: From tools to tissue models. *Curr. Opin. Biotechnol.* **2015**, *35*, 118–126. [CrossRef] [PubMed]

24. Chen, H.C. Boyden Chamber Assay. *Methods Mol. Biol.* **2005**, *294*, 15–22. [CrossRef] [PubMed]

25. Liu, T.; Li, C.; Li, H.; Zeng, S.; Qin, J.; Lin, B. A microfluidic device for characterizing the invasion of cancer cells in 3-D matrix. *Electrophoresis* **2009**, *30*, 4285–4291. [CrossRef] [PubMed]

26. Haessler, U.; Teo, J.C.M.; Fortay, D.; Renaud, P.; Swartz, M.A. Migration dynamics of breast cancer cells in a tunable 3D interstitial flow chamber. *Integr. Biol.* **2012**, *4*, 401–409. [CrossRef] [PubMed]

27. Kubota, Y.; Kleinmann, H.K.; Martin, G.R.; Lawley, T.J. Role of Laminin and Basement Membrane in the Morphological Differentiation of Human Endothelial Cells into Capillary-like Structures. *J. Cell Biol.* **1988**, *107*, 1589–1598. [CrossRef] [PubMed]

28. Bowyer, S.; Kanthou, C.; Reyes-Aldasoro, C.C. Analysis of capillary-like structures formed by endothelial cells in a novel organotypic assay developed from heart tissue. In *Medical Image Understanding and Analysis*, 1st ed.; Reyes-Aldasoro, C.C., Slabaugh, G.G., Eds.; Springer: Cham, Switzerland, 2014; pp. 226–231, ISBN 978-3319609638.

29. Lee, H.; Chung, M.; Jeon, N.L. Microvasculature: An essential component for organ-on-chip systems. *MRS Bull.* **2014**, *39*, 51–59. [CrossRef]

30. Chen, M.B.; Whisler, J.A.; Jeon, J.S.; Kamm, R.D. Mechanism of tumor cell extravasation in an in vitro microvascular network platform. *Integr. Biol.* **2013**, *5*, 1262–1271. [CrossRef] [PubMed]

31. Bersini, S.; Jeon, S.; Dubini, G.; Arrigoni, C.; Chung, S.; Charest, J.L.; Moretti, M.; Kamm, R.D. A Microfluidic 3D in vitro model for specificity of breast cancer metastasis to bone. *Biomaterials* **2014**, *35*, 2454–2461. [CrossRef] [PubMed]

32. Jeon, J.S.; Bersini, S.; Gilardi, M.; Dubini, G.; Charest, J.L.; Moretti, M.; Kamm, R.D. Human 3D vascularized organotypic microfluidic assays to study breast cancer cell extravasation. *PNAS* **2015**, *112*, 214–219. [CrossRef] [PubMed]

33. Shin, M.K.; Kim, S.K.; Jung, H. Integration of intra- and extravasation in one cell-based microfluidic chip for the study of cancer metastasis. *Lab Chip* **2011**, *11*, 3880–3887. [CrossRef] [PubMed]

34. Riahi, R.; Yang, Y.L.; Ki, H.; Jiang, L.; Wong, P.K.; Zohar, Y. A microfluidic model for organ-specific extravasation of circulating tumor cells. *Biomicrofluidics* **2014**, *8*, 024103. [CrossRef] [PubMed]

35. Zervantonakis, J.K.; Hughes-Alford, S.K.; Charest, J.L.; Condeelis, J.S.; Gertler, F.B.; Kamm, R.D. Three-dimensional microfluidic model for tumor cell intravasation and endothelial barrier function. *PNAS* **2012**, *109*, 13515–13520. [CrossRef] [PubMed]

36. Song, J.W.; Cavnar, S.P.; Walker, A.C.; Luker, K.E.; Gupta, M. Microfluidic Endothelium for Studying the Intravascular Adhesion of Metastatic Breast Cancer Cells. *PLoS ONE* **2009**, *4*, e5756. [CrossRef] [PubMed]

37. Zhang, Q.; Liu, T.; Qin, J. A microfluidic-based device for study of transendothelial invasion of tumor aggregates in realtime. *Lab Chip* **2012**, *12*, 2837–2842. [CrossRef] [PubMed]

38. Cui, X.; Guo, W.; Sun, Y.; Sun, B.; Hu, S.; Sun, D.; Lam, R.H.W. A microfluidic device for isolation and characterization of transendothelial migrating cancer cells. *Biomicrofluidics* **2017**, *11*, 014105. [CrossRef] [PubMed]

39. Weinberg, C.B.; Bell, E. A blood vessel model constructed from collagen and cultured vascular cells. *Science* **1986**, *231*, 397–400. [CrossRef] [PubMed]

40. Dejana, E. Endothelial Cell–Cell Junctions: Happy Together. *Nat. Rev. Mol. Cell Biol.* **2004**, *5*, 261–270. [CrossRef] [PubMed]

41. Kramer, R.H.; Bensch, K.G.; Davison, P.M.; Karasek, M.A. Basal Lamina Formation by Cultured Microvascular Endothelial Cells. *J. Cell Biol.* **1984**, *99*, 692–698. [CrossRef] [PubMed]

42. Kühn, K. Basement Membrane (Type IV) Collagen. *Matrix Biol.* **1994**, *14*, 439–445. [CrossRef]

43. Stratman, A.N.; Malotte, K.M.; Maham, R.D.; Davis, M.J.; Davis, G.E. Pericyte recruitment during vasculogenic tube assembly stimulates endothelial basement membrane matrix formation. *Blood* **2009**, *114*, 5091–5101. [CrossRef] [PubMed]

44. Vu, T.H.; Shipley, J.M.; Bergers, G.; Berger, J.E.; Helms, J.A.; Hanahan, D.; Shapiro, S.D.; Senior, R.M.; Werb, Z. MMP-9/Gelatinase B Is a Key Regulator of Growth Plate Angiogenesis and Apoptosis of Hypertrophic Chondrocytes. *Cell* **1998**, *93*, 411–422. [CrossRef]

45. Timpl, R. Structure and biological activity of basement membrane proteins. *Eur. J. Biochem.* **1989**, *180*, 487–502. [CrossRef] [PubMed]

46. Paulsson, M.M. Basement membrane Proteins: Structure, Assembly, and Cellular Interactions. *Crit. Rev. Biochem. Mol. Biol.* **1992**, *27*, 93–127. [CrossRef] [PubMed]

47. Hughes, C.S.; Postovit, L.M.; Lajoie, G.A. Matrigel: A complex protein mixture required for optimal growth of cell culture. *Proteomics* **2010**, *10*, 1886–1890. [CrossRef] [PubMed]

48. Reymond, N.; d'Água, B.B.; Ridley, A.J. Crossing the endothelial barrier during metastasis. *Nat. Rev. Cancer* **2013**, *13*, 858–870. [CrossRef] [PubMed]

49. Chaw, K.C.; Manimaran, M.; Tayad, E.H.; Swaminathan, S. Multi-step microfluidic device for studying cancer metastasis. *Lab Chip* **2007**, *7*, 1041–1047. [CrossRef] [PubMed]

50. Langley, R.R.; Fidler, I.J. The seed and soil hypothesis revisited—The role of tumorstroma interactions in metastasis to different organs. *Int. J. Cancer* **2011**, *128*, 2527–2535. [CrossRef] [PubMed]

51. Fidler, I.J.; Poste, G. The "seed and soil" hypothesis revisited. *Lancet Oncol.* **2008**, *9*, 808. [CrossRef]

52. Attaye, I.; Smulders, Y.M.; de Waard, M.C.; Oudemans-van Straaten, H.M.; Smit, B.; Van Wijhe, M.H.; Musters, R.J.; Koolwijk, P.; Spoelstra-de Man, A.M. The effects of hypoxia on microvascular endothelial cell proliferation and production of vaso-active substances. *Intensive Care Med. Exp.* **2017**, *5*, 22. [CrossRef] [PubMed]

53. Coslett, L.M.; Katlic, M.R. Lung cancer with skin metastasis. *Chest* **1990**, *97*, 757–759. [CrossRef] [PubMed]

54. Ripley, R.T.; Rusch, V.W. Lung metastases. In *Abeloff's Clinical Oncology*, 5th ed.; Niederhuber, J.E., Armitage, J.O., Doroshow, J.H., Kastan, M.B., Tepper, J.E., Eds.; Elsevier Saunders: Philadelphia, PA, USA, 2014; Chapter 52, ISBN 978-1455728657.

55. Popper, H.H. Progression and metastasis of lung cancer. *Cancer Metastasis Rev.* **2016**, *35*, 75–91. [CrossRef] [PubMed]

Bioengineering **2018**, *5*, 40

56. Burton, A.C. Relation of Structure to Function of the Tissues of the Wall of Blood Vessels. *Physiol. Rev.* **1954**, *34*, 619–642. [CrossRef] [PubMed]

57. Dewey, C.F.; Bussolari, S.R.; Gimbrone, M.A.; Davies, P.F. The Dynamic Response of Vascular Endothelial Cells to Fluid Shear Stress. *J. Biomech. Eng.* **1981**, *103*, 177–185. [CrossRef] [PubMed]

58. Levesque, M.J.; Nerem, R.M. The Elongation and Orientation of Cultured Endothelial Cells in Response to Shear Stress. *J. Biomech. Eng.* **1985**, *107*, 341–347. [CrossRef] [PubMed]

59. Adams, D.H.; Shaw, S. Leucocyte-endothelial interactions and regulation of leucocyte migration. *Lancet* **1994**, *343*, 831–836. [CrossRef]

60. Marieb, E.N.; Hoehn, K. The cardiovascular system: Blood vessels. In *Pearson Education*, 9th ed.; Human Anatomy & Physiology: London, UK, 2013; p. 712. ISBN 978-0-321-74326-8.

61. Laferriere, J.; Houle, F.; Taher, M.M.; Valerie, K.; Huot, J. Transendothelial Migration of Colon Carcinoma Cells Requires Expression of E-selectin by Endothelial Cells and Activation of Stress-activated Protein Kinase-2 (SAPK2/p38) in the Tumor Cells. *J. Biol. Chem.* **2001**, *276*, 33762–33772. [CrossRef] [PubMed]

62. Chappel, D.C.; Varner, S.E.; Nerem, R.M.; Medford, R.M.; Alexander, W. Oscillatory Shear Stress Stimulates Adhesion Molecule Expression in Cultured Human Endothelium. *Circ. Res.* **1998**, *82*, 532–539. [CrossRef]

63. Nagrath, S.; Sequist, L.V.; Maheswaran, S.; Bell, D.W.; Irima, D.; Ulkus, L.; Smith, M.R.; Kwal, E.L.; Digumarthy, S.; Muzikansky, A.; et al. Isolation of rare circulating tumour cells in cancer patients by microchip technology. *Nature* **2007**, *450*, 1235–1239. [CrossRef] [PubMed]

bioengineering

MDPI

Article

A 3D Microfluidic Model to Recapitulate Cancer Cell Migration and Invasion

Yi-Chin Toh [1,2], Anju Raja [2,3], Hanry Yu [2,4,5,6,7,8,9] and Danny van Noort [10,11,*]

1   Department of Biomedical Engineering, 4 Engineering Drive, National University of Singapore, Singapore 117853, Singapore; biety@nus.edu.sg
2   Institute of Bioengineering and Nanotechnology, A*STAR, The Nanos, #04-01, 31 Biopolis Way, Singapore 138669, Singapore
3   Integrated Health Information Systems (IHiS), 6 Serangoon North Avenue 5, Singapore 554910, Singapore; anju.mythreyi.raja@ihis.com.sg
4   Department of Physiology, Yong Loo Lin School of Medicine, MD9-04-11, 2 Medical Drive, Singapore 117597, Singapore; medyuh@nus.edu.sg
5   Mechanobiology Institute, National University of Singapore, T-Lab, #05-01, 5A Engineering Drive 1, Singapore 117411, Singapore
6   Singapore-MIT Alliance for Research and Technology, 1 CREATE Way, #10-01 CREATE Tower, Singapore 138602, Singapore
7   NUS Graduate Programme in Bioengineering, NUS Graduate School for Integrative Sciences and Engineering, National University of Singapore, Singapore 117597, Singapore
8   Department of Biological Engineering, Massachusetts Institute of Technology, Cambridge, MA 02139, USA
9   Gastroenterology Department, Southern Medical University, Guangzhou 510515, China
10   Division of Biotechnology, IFM, Linköping University, Linköping 58183, Sweden
11   Department of New Biology, Daegu Gyeongbuk Institute of Science and Technology (DGIST), Daegu 42988, Korea
*   Correspondence: drr.dvn@gmail.com; Tel.: +65-9243-7077

Received: 14 March 2018; Accepted: 4 April 2018; Published: 8 April 2018

Abstract: We have developed a microfluidic-based culture chip to simulate cancer cell migration and invasion across the basement membrane. In this microfluidic chip, a 3D microenvironment is engineered to culture metastatic breast cancer cells (MX1) in a 3D tumor model. A chemo-attractant was incorporated to stimulate motility across the membrane. We validated the usefulness of the chip by tracking the motilities of the cancer cells in the system, showing them to be migrating or invading (akin to metastasis). It is shown that our system can monitor cell migration in real time, as compare to Boyden chambers, for example. Thus, the chip will be of interest to the drug-screening community as it can potentially be used to monitor the behavior of cancer cell motility, and, therefore, metastasis, in the presence of anti-cancer drugs.

Keywords: 3D cell culture; microfluidics; cell migration; cell invasion; metastasis

---

1. Introduction

Metastasis is a leading cause of death in patients with malignant neoplasms [1]. The mechanism of metastasis has been under intense research and may translate into effective cancer therapies [2–4]. Cancer metastasis progresses in multiple steps. It involves the loss of cell adhesion from the primary tumor, increased cell motility and invasion across the basement membrane into the blood capillary (intravasation), systemic circulation and, finally, extravasation into surrounding tissues [5]. The assays for cancer metastatic potential are typically pursued in in vivo models since they have all the necessary cues essential for successful metastasis [6,7]. However, animal models are expensive and difficult to multiplex [8]. Moreover, it is difficult to isolate and study the multi-factorial processes contributing to metastasis. In vitro models allow for more controlled experimentation to better understand specific processes, such as migration and invasion, leading to cancer metastasis [9].

The development of microfluidic systems as in vitro cancer cell migration models, in particular, offers advantages over conventional methods of studying cancer cell migration and invasion, such as Boyden chambers [10–12] and scratch tests [12–14]. Microfluidic-based cancer migration models minimize the requirements for reagents and cells [15]. Such a platform is particularly useful for the study of small cancer cell populations, such as cancer stem cells or cells obtained from clinical patient specimens. Microfluidic cancer migration models also allow the application of chemo-attractant gradients [16,17], improve imaging resolution [18] and can look at interaction with other cells [19,20]. However, cancer cells in these microfluidic cancer migration models are cultured 2-dimensionally, rendering them only suitable for studying the inherent genetic migratory disposition of a cancer cell population. These models lack the context of a 3D tumor microenvironment to investigate the onset and progression of cancer cell migration and invasion, which has been increasingly implicated in cancer metastasis [8,21,22]. Improvements have been achieved by incorporating Matrigel^{TM} lining [18] or collagen scaffolds [23] into microfluidic cell migration models to observe how cancer cells migrate across a 3D barrier. One such a barrier was formed of an endothelial layer in which the intravasation of cancer cells was monitored [24]. It should also be noted that cancer cell density influences the cell migration [25,26]. Therefore, as cell migration starts with solid tumors, it is imperative to include solid cancer cell aggregates in the migration model to study cancer. This has not been the case in most of the previous studies.

Here, we describe a microfluidic cancer cell migration model that allows cancer cells to form 3D cellular aggregates resembling cancer tumors before initiating cancer cell migration and invasion. To mimic the basement membrane, a 3D collagen barrier is then formed around the 3D cancer cell aggregate via a polyelectrolyte complex coacervation process described by Toh et al. [27]. We were able to observe, in real time, the migration and invasion of a metastatic breast cancer cell (MX-1) from a 3D cellular aggregate across a collagen barrier. Our system also has excellent optical properties and allows multi-dimensional (x,y,z, time) acquisition of the cell migration and invasion process at high resolution. Thus, our microfluidic cancer cell migration model presents an opportunity to study cancer cell migration at high spatial and temporal resolution in a more biologically relevant 3D setting.

2. Materials and Methods

We first formed a 3D cancer aggregate by engineering a 3D microenvironment within a 1 cm (length) × 600 μm (width) × 100 μm (height) polydimethylsiloxane (PDMS) microfluidic channel. The fabrication of the microfluidic channel was previously described by Toh et al. [27]. An array of 30 × 50 μm elliptical micropillars with a gap size of 20 μm separated the microfluidic channel into 3 compartments: a 200 μm wide central cell culture compartment flanked by 2 side perfusion compartments. The pillar dimensions and gap size determine the porosity of the pillar array and, therefore, the exposure of the cells towards sheer stress of the perfused medium and the level of diffusion of nutrients and waste across the pillar array. The micropillar array within the microfluidic channel immobilizes cells at high density, forming 3D cell-cell interactions (Figure 1A). After the cells were seeded, a cell-conforming layer of 3D matrix was formed by the laminar flow complex coacervation of a positively-charged modified collagen and negatively-charged acrylate-based terpolymer to present the cells with 3D cell-matrix interactions [27]. Cancer cells cultured in this 3D microfluidic cell culture system remodeled into a 3D cellular aggregate, which exhibited cortical actin localization and expressed the cell adhesion protein E-cadherin (Figure 1B,C), indicating a high cell density with tight junctions. After the formation of a 3D cellular aggregate, we constructed a collagen barrier around the aggregate to simulate the basement membrane which migrating cancer cells must transverse during invasion (Figure 1D). The collagen barrier was formed by using the laminar flow complex coacervation process as described previously. However, the ratio of the negatively charged acrylate-based terpolymer to positively charged modified collagen was kept <5 to ensure a sufficiently thick collagen gel to be formed [27]. To induce cancer cells to migrate from the center cell compartment to the side channels, chemo-attractants were perfused through the two side channels (Figure 1D). Due to the slow diffusion rate of the chemo-attractant across the collagen barrier [28], a gradient was maintained for the duration of the experiments.

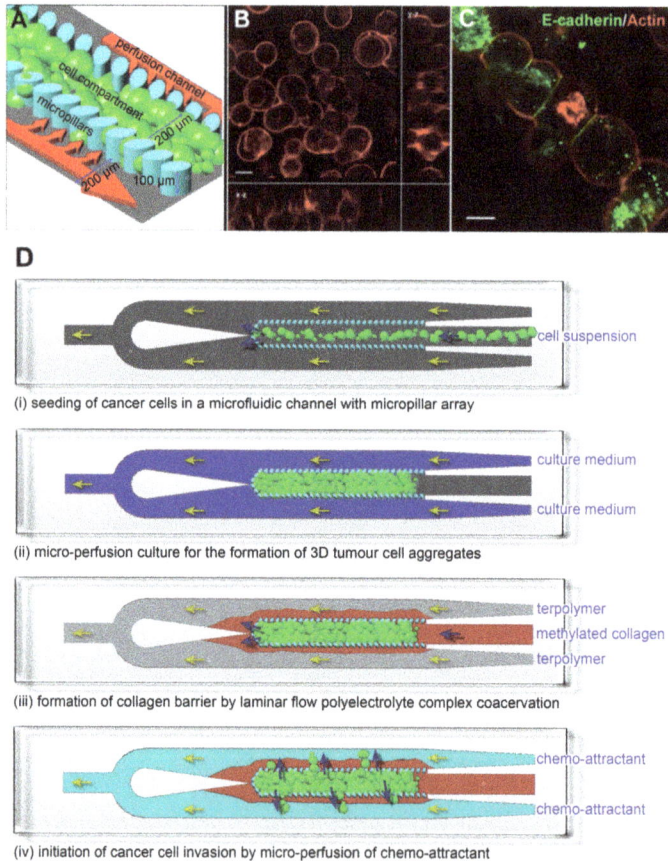

Figure 1. Establishment of a 3D cancer cell migration model in a microfluidic channel. (A) An array of 30×50 μm micropillars separated the microfluidic channel into 3 compartments: a central cell culture compartment and 2 side media perfusion compartments. Cancer cells are immobilized 3-dimensionally at high density within the central cell compartment and will remodel into 3D cellular aggregates after perfusion culture; (B,C) show the 3D cellular phenotype of MCF7, a breast cancer cell line, after 3 days of perfusion culture. (B) Rhodamine-phalloidin staining revealed cortical actin distribution. Image is an orthogonal projection of a 30 μm thick confocal optical section; (C) E-cadherin immunofluorescence staining confirmed the presence of cell-cell interactions within the 3D cellular aggregate. Scale bars = 10 μm; (D) Schematics for performing the cancer cell migration/invasion assay using the microfluidic model. (i) Cancer cells are seeded into the microfluidic channel and (ii) perfusion-cultured for 3 days to allow formation of 3D cellular aggregate. (iii) A collagen barrier is formed around the 3D cellular aggregate by laminar flow complex coacervation of a positively charged collagen and a negatively charged HEMA-MMA-MAA terpolymer. (iv) Cancer cell migration/invasion is then initiated by perfusing chemo-attractant through the side perfusion channels.

We validated our microfluidic cancer cell migration model with a breast cancer cell model. MX-1 cells were routinely maintained in RPMI medium (Invitrogen, Singapore) with 10% fetal calf serum (FCS) (Invitrogen, Singapore), 1.5 g/L sodium bicarbonate (Invitrogen, Singapore), 1 mM sodium pyruvate (Invitrogen, Singapore) and 1.5 g/L L-glutamate (Invitrogen, Singapore) at 37 °C, 5% CO_2. The cells were seeded into the 3D microfluidic cell culture system at a density of 5×10^6 cells/mL and perfusion-cultured at a flow rate of 0.03 mL/h for 3 days to allow formation of 3D MX-1 aggregates. During this period, the cells were serum-starved from 10% to 5% FCS, with a 2.5% reduction in serum concentration every 24 h. Upon formation of the 3D MX-1 aggregates, the collagen barrier was formed around the aggregate. The cell migration assay was initiated by perfusing a chemo-attractant (RMPI medium with 20% FCS and 60 mM HEPES (Invitrogen, Singapore)) through the side channels at a flow rate of 0.02 mL/h. The migration of MX-1 cells from the 3D aggregate across the collagen barrier into the side perfusion channels was captured by time-lapse video (Streampix version 3.12.2, NorPix Inc., Quebec, Montreal, Canada) on a heated stage at 37 °C of a microscope (Olympus, Tokyo, Japan) for 45 h. Two groups (migratory and invading) of 3 cells were tracked by marking their position in the time-lapse video. Their x,y coordinates were then determined in Mathematica (Wolfram Research Inc., Champaign, IL, USA). From the time-lapse video the number of invading cells and cells in the channel section were estimated which gave the percentage of invading cells in the system. The migration rate was calculated as the total distance travelled in 45 h and averaged over the 3 cells in each group. We compared the invasiveness of MX-1 cells in the microfluidic model with that seen in conventional Boyden chambers. For Boyden chamber cell invasion assay, 150,000 cells were seeded in each ECM coated invasion assay chamber (Cell Invasion Assay Kit, Merck Millipore, Darmstadt, Germany) with 5% serum and placed in 24 well plates containing media with 20% serum to act as a chemo-attractant for invasion. Wells were assayed every 12 h, where the cells and matrix inside the wells were removed using cotton swabs and the bottom of the well was immersed in a cell staining solution (Merck Millipore, Darmstadt, Germany) for 20 min, rinsed in water and the number of invasive cells were counted.

3. Results and Discussion

In our microfluidic cancer cell migration model, we observed that MX-1 cells were able to invade across the collagen barrier into the side perfusion channels within 45 h (Figure 2A, Supplementary Video S1). The invading MX-1 cells in the microfluidic model exhibited both amoeboid-like motility, where the cells had amorphous cell morphology and changed direction rapidly and mesenchymal-like motility, where cells are elongated and form membrane protrusions at the leading edge [8,29] (Figure 2A). The amoeboid mode of cancer cell motility is usually only observed in animal or 3D in vitro models, and is mechanistically different from mesenchymal motility observed when cells are cultured 2-dimensionally [8,29]. We were also able to observe collective motility where cells retained their cell-cell contacts and invade as a group (Figure 2A). This mode of cancer cell motility has previously been observed in animal models only and is poorly understood because in vitro models have not been successful in modeling such motility [8,29]. The 3D tumor microenvironment in our microfluidic model provides the necessary cues to allow cancer cells to exhibit different modes of motility as compared to those typically observed in Boyden chambers and other 2D cell migration models. Since the underlying mechanisms for different modes of motility are different and there is plasticity in motility modes, anti-metastatic drugs that are successful in inhibiting mesenchymal cell migration in 2D cell migration models (e.g., Boyden chambers) may not be clinically effective [8,30]. Our system can function as an alternative or complementary testing model to existing cell migration models for evaluating a combination of drug inhibitors targeting different modes of cancer cell motility.

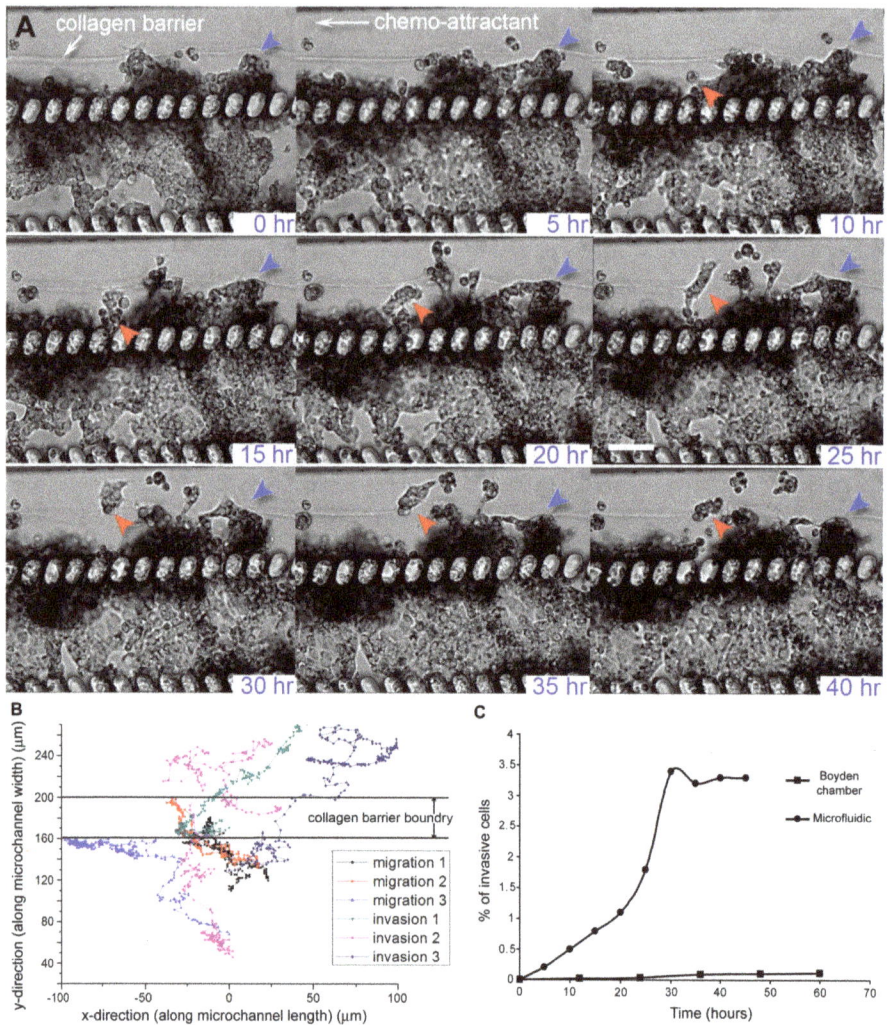

Figure 2. Migration and invasion of MX-1, a metastatic breast cancer cell line, in the microfluidic cancer cell migration model. (**A**) Time-lapse transmission images showing that MX-1 cells exhibited different modes of motility as they migrated or invaded across the collagen barrier over a period of 40 h. Most of the cells migrated collectively instead as single cell (red and blue arrows). Cells also exhibited plasticity in their mode of motility. Cells (in red arrow) displayed mesenchymal-like motility for up to 35 h before switching to amoeboid-like motility. Scale bar = 100 μm; (**B**) Migration trajectories of MX-1 cells showed the presence of 2 cell populations. Cells that transmigrated across the collagen barrier were defined as invading cells while cells that were motile but did not transmigrate across the collagen barrier were defined as migratory cells. The boundary of the collagen barrier varied over a range of 40 μm because migrating cells caused distention of the barrier at localized regions; (**C**) % of invasive MX-1 cells in the microfluidic cell migration model and control collagen-coated Boyden chamber.

While the majority of the metastatic MX-1 cells were highly motile when compared to non-metastatic breast cancer cell lines, such as MCF7 (data not shown), not all cells within the 3D tumor aggregate have equal tendency to invade. Also, other cancer cell lines, such as liver, kidney, lung, or beta cells, have not been observed to migrate within the presented system [27,31–33].

MX-1 cells that transmigrated across the collagen barrier were defined as invading cells while cells that were motile but did not transmigrate across the collagen barrier were defined as migratory cells. The presence of 2 distinct migrating and invading cell populations was more apparent when we tracked the trajectories of different cells (Figure 2B). The average velocities of the migrating and invading cells were 6.6 ± 1.5 μm/h and 13.5 ± 5.5 μm/h respectively. This observation supports the existence of a heterogeneous cell population within a tumor, where some cells have a higher tendency to intravasate into blood capillaries, thus increasing the likelihood of cancer metastasis [34]. Moreover, when we compared the percentage of cell invasion in the microfluidic model and collagen-coated Boyden chamber, we found that the percentage of invaded cells in the microfluidic model was approximately 10-fold higher than that of the Boyden chamber (Figure 2C). The number of invasive cells will stay constant after a stable gradient condition of the chemo-attractant has been reached. There is increasing evidence showing that a more invasive phenotype is a result of not only intrinsic genetic variation but also a different tumor microenvironment (i.e., ECM and surrounding cells) [5,34]. Cancer cell invasion assays that use in vitro cancer models where cells are cultured on 2D substrates are unable to model the tumor microenvironment adequately to account for the effect of the microenvironment on the invasiveness of cancer cells. In comparison, a 3D tumor microenvironment is represented in our microfluidic cancer invasion model by the construction of 3D tumor aggregate and 3D collagen matrix. Thus, our microfluidic model can facilitate experiments to understand the role of the tumor microenvironment in the development of heterogeneity and invasiveness of cancer cell phenotypes.

In vivo cell migration occurs dynamically in 3D space. To fully study this complex process, not only must an in vitro model recapitulate the 3D tumor microenvironment, it must also be able to capture this process in multiple dimensions (x,y,z, time) at high resolution [35]. The microfluidic cell migration model is designed such that the cellular phenotype and migration trajectory can be imaged and captured at high resolution. The imaging plane of our microfluidic model is perpendicular to the ECM barrier through which cancer cells transmigrate. This design facilitates high-resolution imaging of the migration process as compared to conventional Boyden chambers where the imaging plane is parallel to the ECM barrier [8,36]. In Figure 2, we demonstrated the use of transmission time-lapse microscopy to capture the migration of MX1 cells across the collagen barrier. The microfluidic cell migration model is also amenable to imaging by high-resolution imaging modalities such as laser scanning confocal microscopy, since it is bonded onto glass coverslips. This allows improved spatial resolution and the acquisition of 3D information. We demonstrate this by using laser scanning confocal microscopy to image fluorescent labeled MX-1 cells and collagen matrix in multi-dimensions (x,y,z, time). MX-1 cells were GFP-labeled while the methylated collagen used for forming the collagen barrier was labeled with Alexa Fluor 532 (Invitrogen, Waltham, MA USA) [27]. MX-1 cells were seeded into the microfluidic channel and initiated to migrate as described above. The microfluidic system was placed on a heated stage of a confocal microscope (Zeiss LSM, Jena, Germany) with chemo-attractant (RMPI with 20% FCS and 0.06 mM HEPES (Invitrogen, Singapore) being perfused constantly at a flow rate of 0.02 mL/h. An optical stack of 100 μm (dz = 10 μm) was imaged over 24 h.

MX-1 cells and the collagen matrix in the microfluidic channel can be resolved more clearly by fluorescence confocal imaging than transmission imaging (Figure 3A). The 3D MX-1 aggregate was encapsulated by a fibrous collagen matrix approximately 50 μm thick. The thickness of this collagen barrier is comparable to that of commercially available ECM-coated membranes used in invasion assays with Boyden chambers. Compatibility of our microfluidic cell migration model with fluorescence imaging potentially allows different cell populations in a tumor aggregate, such as cancer-initiating and non-initiating cells [37], to be differentially labeled and tracked during the intravasation process. 3D information can be extracted by performing a 3D reconstruction of the model as shown in Figure 3B or by performing image processing of the individual sections of the optical stack. As observed using transmission microscopy, invasion of MX-1 cells across the collagen barrier occurred at localized regions (Figure 3C). Confocal microscopy revealed that invasion was initiated by the establishment of contact between migrating cells and the collagen matrix at a localized spot, followed by degradation of the collagen matrix over a period of 24 h (Figure 3C). Our microfluidic cancer migration model was able to clearly distinguish hallmark events of cancer intravasation (i.e., increased cell motility, diminished cell adhesions and proteolytic disruption of the ECM) [5] at real time in 3D space. When compared to models that assay for end points such as number of cells migrated [18,38], our microfluidic model shows more versatility for screening anti-metastatic drugs specifically targeting a process of intravasation. For example, we can evaluate the efficacy of anti-metastatic drugs targeting at metalloproteinases (MMPs) [39] by tracking the degradation of the collagen matrix over time. Alternatively, measuring changes in the rate of cell migration can help to evaluate the effectiveness of drugs targeting cell motility [40].

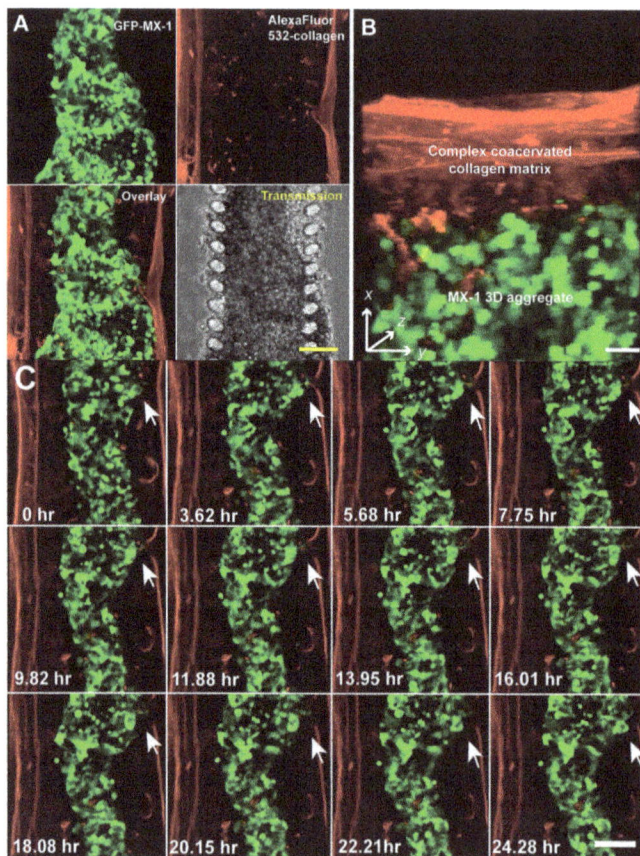

Figure 3. Multi-dimensional (x, y, z, time) imaging of cancer cell migration in the microfluidic cell migration model using a laser scanning confocal microscope. (**A**) MX-1 cells and collagen matrix were labeled with GFP and Alexa Fluor 532 respectively so that they can be imaged independently. Scale bar = 100 μm; (**B**) 3D reconstruction of a 10 μm optical section at 1 μm interval. Scale bar = 50 μm; (**C**) 24 h time-lapse confocal imaging showing remodeling of the collagen matrix by invading MX-1 cells (white arrow). Scale bar = 100 μm.

Bioengineering **2018**, *5*, 29

4. Conclusions

In conclusion, we developed a microfluidic cell migration model that can resolve different aspects of cancer cell intravasation (i.e., loss of cell adhesion, different modes of cell motility and ECM degradation) at high resolution in a biologically relevant 3D environment. We were able to accomplish this by incorporating a 3D microenvironment, which has been previously shown to affect the invasiveness of cancer cells, into a transparent microfluidic system. This microfluidic cancer cell migration model not only has high imaging resolution and resemblance to the in vivo situation, but it is also amenable to multiplexing to achieve high-throughput assaying. Hence, our microfluidic cancer cell migration model is appealing for anti-metastasis drug testing, especially drugs targeting migration and invasion.

Supplementary Materials: The following are available online at http://www.mdpi.com/2306-5354/5/2/29/s1, Video S1: MX1_0_48hr.avi.

Acknowledgments: This work is supported in part by the Institute of Bioengineering and Nanotechnology, Biomedical Research Council, Agency for Science, Technology and Research (A*STAR), A*STAR, (Project Number 1334i00051); NMRC (R-185-000-294-511); NUHS Innovation Seed Grant 2017 (R-185-000-343-733); MOE ARC (R-185-000-342-112); SMART BioSyM and Mechanobiology Institute of Singapore (R-714-006-008-271) funding to HYU. We are grateful to Dr. Robert Hoffman of AntiCancer Inc. for providing us with the GFP-labeled MX-1 cells.

Author Contributions: Yi-Chin Toh did most of the experiments and the paper writing, as well as experimental design. Anju Raja performed parts of the experiments. Hanry Yu conceived and partly designed the experiment. Danny van Noort did the image analysis and tracking data as well as part of the paper writing and revision.

Conflicts of Interest: The authors declare no conflict of interest.

References

1. Chaffer, C.L.; Weinberg, R.A. A Perspective on Cancer Cell Metastasis. *Science* **2011**, *331*, 1559–1564. [CrossRef] [PubMed]
2. Duffy, M.J.; McGowan, P.M.; Gallagher, W.M. Cancer invasion and metastasis: Changing views. *J. Pathol.* **2008**, *214*, 283–293. [CrossRef] [PubMed]
3. Zijlstra, A.; Lewis, J.; DeGryse, B.; Stuhlmann, H.; Quigley, J.P. The Inhibition of Tumor Cell Intravasation and Subsequent Metastasis via Regulation of In Vivo Tumor Cell Motility by the Tetraspanin CD151. *Cancer Cell* **2008**, *13*, 221–234. [CrossRef] [PubMed]
4. Eckhardt, B.L.; Francis, P.A.; Parker, B.S.; Anderson, R.L. Strategies for the discovery and development of therapies for metastatic breast cancer. *Nat. Rev. Drug Discov.* **2012**, *11*, 479–497. [CrossRef] [PubMed]
5. Gupta, G.P.; Massagué, J. Cancer metastasis: Building a framework. *Cell* **2006**, *127*, 679–695. [CrossRef] [PubMed]
6. Paweletz, C.P.; Charboneau, L.; Liotta, L.A. Overview of Metastasis Assays. *Curr. Protoc. Cell Biol.* **2001**, *12*, 19.1.1–19.1.9. [CrossRef]
7. Deryugina, E.; Quigley, J. Chick embryo chorioallantoic membrane model systems to study and visualize human tumor cell metastasis. *Histochem. Cell Biol.* **2008**, *130*, 1119–1130. [CrossRef] [PubMed]
8. Sahai, E. Mechanisms of cancer cell invasion. *Curr. Opin. Genet. Dev.* **2005**, *15*, 87–96. [CrossRef] [PubMed]
9. Kenny, H.A.; Krausz, T.; Yamada, S.D.; Lengyel, E. Use of a novel 3D culture model to elucidate the role of mesothelial cells, fibroblasts and extra-cellular matrices on adhesion and invasion of ovarian cancer cells to the omentum. *Int. J. Cancer* **2007**, *121*, 1463–1472. [CrossRef] [PubMed]
10. Chien, W.; O'Kelly, J.; Lu, D.; Leiter, A.; Sohn, J.; Yin, D.; Karlan, B.; Vadgama, J.; Lyons, K.M.; Koeffler, H.P. Expression of connective tissue growth factor (CTGF/CCN2) in breast cancer cells is associated with increased migration and angiogenesis. *Int. J. Oncol.* **2011**, *38*, 1741–1747. [CrossRef] [PubMed]
11. Yamauchi, A.; Yamamura, M.; Katase, N.; Itadani, M.; Okada, N.; Kobiki, K.; Nakamura, M.; Yamaguchi, Y.; Kuribayashi, F. Evaluation of pancreatic cancer cell migration with multiple parameters in vitro by using an optical real-time cell mobility assay device. *BMC Cancer* **2017**, *17*, 234. [CrossRef] [PubMed]
12. Yarrow, J.C.; Perlman, Z.E.; Westwood, N.J.; Mitchison, T.J. A high-throughput cell migration assay using scratch wound healing, a comparison of image-based readout methods. *BMC Biotechnol.* **2004**, *4*, 21. [CrossRef] [PubMed]

13. Xia, M.; Yao, L.; Zhang, Q.; Wang, F.; Mei, H.; Guo, X.; Huang, W. Long noncoding RNA HOTAIR promotes metastasis of renal cell carcinoma by up-regulating histone H3K27 demethylase JMJD3. *Oncotarget* **2017**, *8*, 19795–19802. [CrossRef] [PubMed]

14. Hulkower, K.I.; Herber, R.L. Cell Migration and Invasion Assays as Tools for Drug Discovery. *Pharmaceutics* **2011**, *3*, 107–124. [CrossRef] [PubMed]

15. Nie, F.-Q.; Yamada, M.; Kobayashi, J.; Yamato, M.; Kikuchi, A.; Okano, T. On-chip cell migration assay using microfluidic channels. *Biomaterials* **2007**, *28*, 4017–4022. [CrossRef] [PubMed]

16. Saadi, W.; Wang, S.-J.; Lin, F.; Jeon, N.L. A parallel-gradient microfluidic chamber for quantitative analysis of breast cancer cell chemotaxis. *Biomed. Microdevices* **2006**, *8*, 109–118. [CrossRef] [PubMed]

17. Jeon, N.L.; Baskaran, H.; Dertinger, S.K.W.; Whitesides, G.M.; van der Water, L.; Toner, M. Neutrophil chemotaxis in linear and complex gradients of interleukin-8 formed in a microfabricated device. *Nat. Biotechnol.* **2002**, *20*, 826–830. [CrossRef] [PubMed]

18. Chaw, K.C.; Manimaran, M.; Tay, E.H.; Swaminathan, S. Multi-step microfluidic device for studying cancer metastasis. *Lab Chip* **2007**, *7*, 1041–1047. [CrossRef] [PubMed]

19. Businaro, L.; de Ninno, A.; Schiavoni, G.; Lucarini, V.; Ciasca, G.; Gerardino, A.; Belardelli, F.; Gabriele, L.; Mattei, F. Cross talk between cancer and immune cells: Exploring complex dynamics in a microfluidic environment. *Lab Chip* **2013**, *13*, 229–239. [CrossRef] [PubMed]

20. Mi, S.; Du, Z.; Xu, Y.; Wu, Z.; Qian, X.; Zhang, M.; Sun, W. Microfluidic co-culture system for cancer migratory analysis and anti-metastatic drugs screening. *Sci. Rep.* **2016**, *6*, 35544. [CrossRef] [PubMed]

21. Bogenrieder, T.; Herlyn, M. Axis of evil: Molecular mechanisms of cancer metastasis. *Oncogene* **2003**, *22*, 6524–6536. [CrossRef] [PubMed]

22. Erler, J.T.; Weaver, V.M. Three-dimensional context regulation of metastasis. *Clin. Exp. Metastasis* **2009**, *26*, 35–49. [CrossRef] [PubMed]

23. Chung, S.; Sudo, R.; Mack, P.J.; Wan, C.-R.; Vickerman, V.; Kamm, R.D. Cell migration into scaffolds under co-culture conditions in a microfluidic platform. *Lab Chip* **2009**, *9*, 269–275. [CrossRef] [PubMed]

24. Zervantonakis, I.K.; Hughes-Alford, S.K.; Charest, J.L.; Condeelis, J.S.; Gertler, F.B.; Kamm, R.D. Three-dimensional microfluidic model for tumor cell intravasation and endothelial barrier function. *Proc. Natl. Acad. Sci. USA* **2012**, *109*, 13515–13520. [CrossRef] [PubMed]

25. Choi, Y.; Hyun, E.; Seo, J.; Blundell, C.; Kim, H.C.; Lee, E.; Lee, S.H.; Moon, A.; Moon, W.C.; Huh, D. A microengineered pathophysiological model of early-stage breast cancer. *Lab Chip* **2015**, *15*, 3350–3357. [CrossRef] [PubMed]

26. Chen, Y.; Gao, D.; Liu, H.; Lin, S.; Jiang, Y. Drug cytotoxicity and signaling pathway analysis with three-dimensional tumor spheroids in a microwell-based microfluidic chip for drug screening. *Anal. Chim. Acta* **2015**, *898*, 85–92. [CrossRef] [PubMed]

27. Toh, Y.C.; Zhang, C.; Zhang, J.; Khong, Y.M.; Chang, S.; Samper, V.D.; van Noort, D.; Hutmacher, D.W.; Yu, H. A novel 3D mammalian cell perfusion-culture system in microfluidic channels. *Lab Chip* **2007**, *7*, 302–309. [CrossRef] [PubMed]

28. Kihara, T.; Ito, J.; Miyake, J. Measurement of Biomolecular Diffusion in Extracellular Matrix Condensed by Fibroblasts Using Fluorescence Correlation Spectroscopy. *PLoS ONE* **2013**, *8*, e82382. [CrossRef] [PubMed]

29. Yamazaki, D.; Kurisu, S.; Takenawa, T. Regulation of cancer cell motility through actin reorganization. *Cancer Sci.* **2005**, *96*, 379–386. [CrossRef] [PubMed]

30. Croft, D.R.; Olson, M.F. Regulating the conversion between rounded and elongated modes of cancer cell movement. *Cancer Cell* **2008**, *14*, 349–351. [CrossRef] [PubMed]

31. Ong, S.M.; Zhang, C.; Toh, Y.-C.; Kim, S.-H.; Foo, H.-L.; Tan, C.-H.; van Noort, D.; Park, S.; Yu, H. A gel-free 3D microfluidic cell culture system. *Biomaterials* **2008**, *29*, 3237–3244. [CrossRef] [PubMed]

32. Zhang, C.; Zhao, Z.; Abdul Rahim, N.A.; van Noort, D.; Yu, H. Towards human on a chip: Culturing multiple cell types on a chip with compartmentalized microenvironments. *Lab Chip* **2009**, *9*, 3185–3192. [CrossRef] [PubMed]

33. Nguyen, D.T.T.; van Noort, D.; Jeong, I.K.; Park, S. Endocrine systems on chip for a diabetes treatment model. *Biofabrication* **2017**, *9*, 015021. [CrossRef] [PubMed]

34. Isaiah, J.F. The organ microenvironment and cancer metastasis. *Differentiation* **2002**, *70*, 498–505. [CrossRef]

35. Brandt, B.; Heyder, C.; Gloria-Maercker, E.; Hatzmann, W.; Rötger, A.; Kemming, D.; Zänker, K.S.; Entschladen, F.; Dittmar, T. 3D-extravasation model—Selection of highly motile and metastatic cancer cells. *Sem. Cancer Biol.* **2005**, *15*, 387–395. [CrossRef] [PubMed]

36. Heyder, C.; Gloria, M.; Gloria-Maercker, E.; Hatzmann, W.; Niggermann, D.; Zänker, K.S.; Dittmar, T. Realtime visualization of tumor cell/endothelial cell interactions during transmigration across the endothelial barrier. *J. Cancer Res. Clin. Oncol.* **2002**, *128*, 533–538. [CrossRef] [PubMed]

37. Sheridan, C.; Kishimoto, H.; Fuchs, R.; Mehrotra, S.; Bhat-Nakshatri, P.; Turner, C.H.; Goulet, R., Jr.; Badve, S.; Nakshatri, H. CD44+/CD24− breast cancer cells exhibit enhanced invasive properties: An early step necessary for metastasis. *Breast Cancer Res.* **2006**, *8*, R59. [CrossRef] [PubMed]

38. Brown, N.S.; Bicknell, R. Cell Migration and the Boyden Chamber. In *Metastasis Research Protocols*; Brooks, S.A., Schumacher, U., Eds.; Humana Press: Totowa, NJ, USA, 2001; pp. 47–54, ISBN 0-89603-610-3.

39. Overall, C.M.; Kleifeld, O. Tumour microenvironment—Opinion: Validating matrix metalloproteinases as drug targets and anti-targets for cancer therapy. *Nat. Rev. Cancer* **2006**, *6*, 227–239. [CrossRef] [PubMed]

40. Sliva, D. Signaling Pathways Responsible for Cancer Cell Invasion as Targets for Cancer Therapy. *Curr. Cancer Drug Targets* **2004**, *4*, 327–336. [CrossRef] [PubMed]

![bioengineering logo] *bioengineering*

MDPI

Article

Biocatalyst Screening with a Twist: Application of Oxygen Sensors Integrated in Microchannels for Screening Whole Cell Biocatalyst Variants

Ana C. Fernandes [1], Julia M. Halder [2], Bettina M. Nestl [2], Bernhard Hauer [2], Krist V. Gernaey [1] and Ulrich Krühne [1,*]

[1]   Process and Systems Engineering Center (PROSYS), Department of Chemical and Biochemical Engineering, Technical University of Denmark, Building 229, 2800 Kgs. Lyngby, Denmark; ancafe@kt.dtu.dk (A.C.F.); kvg@kt.dtu.dk (K.V.G.)
[2]   Institute of Biochemistry and Technical Biochemistry, Universitaet Stuttgart, 70569 Stuttgart, Germany; julia.halder@itb.uni-stuttgart.de (J.M.H.); bettina.nestl@itb.uni-stuttgart.de (B.M.N.); bernhard.hauer@itb.uni-stuttgart.de (B.H.)
*   Correspondence: ulkr@kt.dtu.dk; Tel.: +45-23-65-21-60

Received: 28 February 2018; Accepted: 5 April 2018; Published: 9 April 2018

Abstract: Selective oxidative functionalization of molecules is a highly relevant and often demanding reaction in organic chemistry. The use of biocatalysts allows the stereo- and regioselective introduction of oxygen molecules in organic compounds at milder conditions and avoids the use of complex group-protection schemes and toxic compounds usually applied in conventional organic chemistry. The identification of enzymes with the adequate properties for the target reaction and/or substrate requires better and faster screening strategies. In this manuscript, a microchannel with integrated oxygen sensors was applied to the screening of wild-type and site-directed mutated variants of naphthalene dioxygenase (NDO) from *Pseudomonas* sp. NICB 9816-4. The oxygen sensors were used to measure the oxygen consumption rate of several variants during the conversion of styrene to 1-phenylethanediol. The oxygen consumption rate allowed the distinguishing of endogenous respiration of the cell host from the oxygen consumed in the reaction. Furthermore, it was possible to identify the higher activity and different reaction rate of two variants, relative to the wild-type NDO. The meander microchannel with integrated oxygen sensors can therefore be used as a simple and fast screening platform for the selection of dioxygenase mutants, in terms of their ability to convert styrene, and potentially in terms of substrate specificity.

Keywords: whole cell biocatalysis; biocatalyst screening; microfluidics; oxygen sensors; dioxygenases; organic chemistry

---

1. Introduction

The use of biocatalysts in industry is increasingly relevant, enabling the production of (new) compounds, some through the conversion of non-natural substrates [1]. Biocatalysts allow stereo- and/or regioselectivity at milder conditions (lower temperatures, close to neutral pH and atmospheric pressure). This is achieved in an often simplified process (remove the need for protection of functional groups in some reactions) that uses less toxic substrates than the equivalent (when existent) organic chemistry methods [1]. However, they can represent a significant fraction of the bioprocess operation cost, due to development costs, need for co-factors, loss of productivity, or complex downstream processing emerging from a low enantioselectivity [2].

Nowadays, around 60% of the industrial biocatalysis is performed using whole cell catalysts [3]. Whole cell biocatalysis solves some of the hurdles of using isolated enzymes, such as catalyst stability in the reaction mixture, enzyme coupling in the cascade reaction system and the need for co-factor addition, by encapsulating the target enzymes in a microbial cell [3,4]. The enzymes involved in the reaction can be from the organism itself but in most cases are from a heterologous source and are inserted in an expression host which is easier to grow, better characterized, considered safe and/or already used in industry (e.g., *Escherichia coli, Pseudomonas* sp., *Trichoderma reesei*) [1,5]. In whole cell biotransformations the main parameters to be optimized for process implementation are the oxygen supply, substrate and product toxicity, product stability and co-factor recycling [6]. All the above-mentioned parameters need to be fully characterized and optimized in order to achieve a cost-effective and productive bioprocess [2]. This requires the selection and/or tailoring of the biocatalyst and/or expression host to the desired reaction or process, as well as the optimization of the reaction and process operating conditions through appropriate screening strategies.

Biocatalyst screening, especially for new processes or products or when involving non-natural substrates, is a complex task, involving different levels of screening [7]. Characteristics related to the final process strategy, such as (i) the target reaction or product; (ii) the type of product (intermediate compound, precursor or the final compound); (iii) the industry (e.g., inorganic chemistry, agrochemical, pharmaceutical, food); (iv) the required degree of purity or enantioselectivity desired; and (v) type of application (for profit or for bioremediation, for example) need to be considered and tested during the screening phase in order to find the optimal combination.

Methods for biocatalyst screening and variant selection require proper design in order to maintain the association between the phenotype observed and the genotype measured [8]. The screening of biocatalysts from initially wide collections, requires the development of both high-throughput growth systems (such as the one developed by Doig et al. (2002) using a robotic liquid handling system [9]) and online quantification and/or monitoring systems.

1.1. Screening Approach

The majority of traditional screening approaches are performed in vitro, thus requiring extraction of the sample from the reaction mixture, or to stop the reaction to perform the quantification of the reaction components. On the other hand, in vivo screening (when the biotransformation/bioconversion is monitored in the microorganisms) broadens the number of parameters that can be screened and accelerates biocatalyst development and selection. However, in vivo screening strategies present some limitations associated with the host's transformation efficiency and growth rate, but also the expression of the target enzyme, limitations in substrate uptake and intracellular background [10].

Most assays tend to be specific (to a certain compound or property, e.g., enatioselectivity [11]) to adequately screen for the intended characteristic. They involve standard analytical equipment such as mass spectrometry (MS), gas chromatography (GC), high-performance liquid chromatography (HPLC) or nuclear magnetic resonance (NMR), thus also being quite time-consuming if hundreds or more samples need to be analyzed. This is mainly due to extra steps in sample preparation such as extraction of the compounds, removal of organic phase and chromatographic separation. This tends to increase both the overall analysis time per sample [12], as well as the amount of potential operator errors.

Faster screening methods often involve the formation or consumption of a colorimetric, luminescent or fluorescent compound, which intensity is related with the substrate's affinity or turnover rate [13]. Spectrophotometric tests are quite desirable due to the possibility of parallelization using microtiter plates, but also owing to the often short measurement time and relatively high sensitivity (signal amplification methods can also be applied). However, most target compounds are neither colorimetric nor fluorescent and developing an indirect colorimetric detection method can be difficult and time-consuming [8]. Moreover, these methods, especially colorimetric methods, can have a high incidence of false positive or negative results which lead to lower precision in the quantification of the target compound [12]. Of the different colorimetric, luminescent or fluorescent methods available that allow long-term measurements, (micro) fluorescence-activated cell sorting ((μ)FACS) devices have been used to screen for growth on a new substrate and to select variants based on enantioselectivity by linking cell survival with capacity for catalysis of only one enantiomer (selection approach) [14–16]. Cell survival can also be linked to antibiotic resistance [8]. Flow cytometry, on the other hand, allows assessing whole cell biocatalysts viability and electron transport, besides enzyme activity, achieving a very high throughput (10^8 screened variants per day [8]).

Another screening method using reporters allows for an indirect measure of the characteristic target, not by the product or substrate of the bioconversion, but by the genetic reporter that encodes that phenotype. This approach is regarded as reaction independent, and the reporter may be colorimetric, fluorescent, bioluminescent or result in conditional survival, cell motility, acidification, or cell display. The activity of the measured reporter is connected to the activity of the target enzyme by interference at either the transcription (e.g., activation of a natural or synthetic transcriptional regulator by binding of the product or substrate), translation (e.g., by binding of the product to a ribozyme or reporter inactivation by the enzyme) or post-translational modification level (e.g., direct modification of the reporter by the enzyme), or even enzyme degradation or solubility (e.g., by fusing green fluorescent protein (GFP) to the enzyme variant, and thus only soluble GFP variants are positively selected). Reporter-based screening using natural transcriptional regulators are usually extremely selective for the target product, having no false positives, but still requires a case-based choice of reporter and approach [10].

1.2. Screening Technologies

A summary of the commonly used screening technologies is presented in Figure 1.

Screening in agar plates usually occurs by formation or disappearance of color on the agar surrounding the incubated colonies or by color appearance on the colonies themselves due to bioconversion, providing a straightforward identification of positive colonies. Differences between activity or catalytic rates are, however, not easily achieved with this type of screening and usually up to 10^5 variants can be analyzed [10]. Screening of 960 variants from cultures grown on a nylon membrane on top of agar plates has been achieved by associating greater intensity of a colored compound with higher producing variants [17]. Coupling of this method with analytical equipment such as HPLC, GC or MS can also be performed. For example, Yan et al. (2017) coupled ambient MS with desorption electrospray ionization (DESI) to achieve a label-free platform for real-time screening of biotransformations in bacterial cultures with imaging MS capable of relating the detected chemical compounds on the surface of the agar plate with their spatial distribution [12]. Metagenomic screening allows the selection of genes with the desired function, by direct phenotypical detection, heterologous complementation, and induced gene expression, usually through isolation in agar plates supplemented with different substrates (e.g., target substrate, certain antibiotics, co-substrates). In this technique, target activity can be further associated with the expression of a reporter gene (e.g., GFP or β-galactosidase) for easier screening [1].

Microtiter plates are the most frequently used screening devices, due to high-throughput and compatibility with a wide range of analytical techniques, from colorimetric approaches to GC, NMR or MS. Microtiter plates are, however, usually limited to libraries of up to 10^4 variants [10]. UV-Vis and fluorescent microplate readers and fluorescent digital imaging (in which mutants are passed on to a nitrocellulose membrane from the agar plate where they were cultured) [18] can be used to increase throughput during biocatalyst selection. Schwaneberg et al. (1999) developed a high-throughput assay using a robotic workstation and spectral analyzer compatible with 96-well microtiter plates for the analysis of substrate specificity and the activity of mutants of a fatty acid hydroxylating enzyme for both pure enzymes and cell extracts [19]. This method was then further developed to allow measurements directly from the whole cell biocatalysts, eliminating the cell lysis, resuspensions and centrifugation preparation steps, and using a replicator tool (developed by Duetz et al. (2000) [20]) to transfer cells from agar plates to the 96-wells plate. Despite the increase in the number of samples analyzed in parallel (~3000 clones per day) as well as directly from the cell bioconversion, issues of reproducibility between screens and differences in activity relative to the one measured in shake flasks (due to evaporation, a well-known problem in open microtiter plates) were observed [21]. Samorski et al. (2005) developed a system for evaluating induction time and growth differences of whole cell catalysts cultured in 96-well microtiter plates with an integrated optical fiber bundle in a x-y-stage. Measurements of optical density (with light scattering and NADH measurements) and amount of expression of an induced fluorescent protein were performed in all wells, allowing the optimization of the initial cell density and the time of induction, and the online monitoring of product formation in a microtiter plate [22]. Codexis, a company specialized in enzyme engineering, and other enzyme developing companies (e.g., Ingenza (Roslin, UK), Nzomics (Newcastle upon Tyne, UK), InnoSyn (Geleen, The Netherlands), Almac (Craigavon, UK)), provide systems for enzyme screening based on 96-well microtiter plates for ketoreductases, acylases, ene-reductases, transaminases, lyases, dehydrogenases and halohydrin dehalogenases [23].

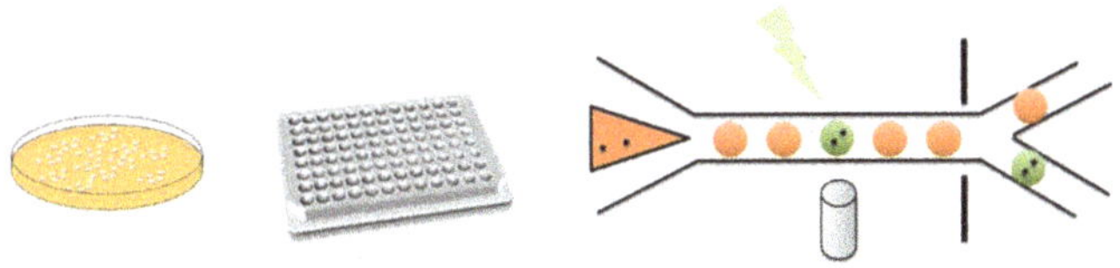

	Agar plate	Microtiter plate	FACS	Droplet microfluidics
Type of sample	Solid; Individual colonies;	Liquid; Sets of cell variants;	Liquid; Sets of cell variants in droplets or Individual cells;	
Library size	10^2 - 10^5	10^4	10^8	10^9
Advantages	Simple; Visual detection by direct phenotypical detection, heterologous complementation and induced gene expression;	Compatible with multiple analytical techniques; High sensitivity; Large dynamic range; Compatible with robotic workstations;	Pico- to femtolitre scale; Low sample and reagent consumption; High-throughput; Continuous operation; Can perform automated sorting;	Compatible with multiple analytical techniques and sensor types; Can perform automated directed evolution;
Disadvantages	Low dynamic range; Labour intensive; Usually requires another method (e.g. microtiter plates) for quantification;	Labour intensive; Works with lysed cells; Small library size; Might require quantification with time-consuming analytics (e.g. GC, MS);	Technnically challenging Limited to fluorescent detection or cell characteristics measured by light scattering;	Requires optimization of emulsions or polyelectrolyte shells used;

Figure 1. Summary of some of the advantages and disadvantages of the commonly used screening approaches for whole cell biocatalysts. GC: gas chromatography; MS: mass spectrometry.

Microfluidic Screening Technologies

Microfluidic approaches for the screening of biocatalysts have emerged recently. Microfluidics, due to the ease of flow manipulation and sensor integration, can potentially enable faster measurement and/or the monitoring of a higher number of parameters [24–26]. Furthermore, microfluidic systems provide continuous production at lower costs and reagent quantities, as well as offering higher safety and less waste generation. They also allow a better spatial and temporal control of the reactions, and operation at unusual process conditions and catalyst configurations, thus expanding the operation conditions and types of reactions possible [27]. FACS on a chip, for example, provides a better control than regular FACS on the number of cells per droplet (down to single cell), as well as on the number of fluorescent reporters per droplet thus presenting highly quantitative results [8]. Abate et al. (2010) used droplet microfluidics to screen mutants of horseradish peroxidase in yeast cells generated by directed evolution, at rates of thousands per second, being 1000-fold faster than microtiter plate-based robotic screening. In this technique, the reaction was initiated on chip, continued in an incubation section, and the cell containing droplets were dielectrophoretically sorted based on fluorescent intensity using a laser connected to a photomultiplier tube. The microfluidic device was designed to allow reinjection of droplets in the system to allow incubation of low activity mutants for later analysis. The detection limit in this case was of ~3500 molecules in 6 pL or <1 turnover per enzyme [28]. Kintses et al. (2012) has applied a similar strategy, but performing fluorescent detection of enzyme reactions in lysates of single cells of *Escherichia coli* (*E. coli*). The best variants were selected with laser-induced fluorescence in a second device with a dielectrophoretic sorter and broken to isolate their plasmid content and use it for the next evolution cycle. Droplets generated this way, where the target compound can be optically read, can also be screened using FACS. The whole cycle of variants' screening developed by Kintses et al. (2012) was performed in two days with the analysis of 10^7 droplets every 3 h [29]. This technique also enables the detection of very small improvements of enzymatic activity [10]. Alternatively, the cells themselves can be used as femtoliter screening vessels, by performing the target reactions inside the cells and detecting the resulting fluorescent product also inside the cells by, for example, FACS. The enzymes can, on the other hand, be displayed on the surface of the cells during the assay by fusion with anchor motifs, being thus more accessible to the substrate (cell surface display technique), but the displayed enzymes may lose activity, and fluorescent substrate or products need to remain bound to the cell surface during measurement. In both approaches, libraries of up to 10^9 variants can be screened [10]. Droplet-based directed evolution of whole cell catalysts is a very powerful tool to increase screening throughput of engineered biocatalysts, but the use of different sensor technologies (such as near infrared (NIR) [30] or Stroboscopic Epifluorescence Imaging [31], for example) is required to expand its application. Further coupling of such a system with standard analytical equipment such as HPLC, GC or MS would increase its usefulness for biotechnologists and the range of detected compounds. Packer and Liu (2015) provide a good overview of available screening methods for protein selection, as well as a short guide on screening method selection [8].

1.3. Screening of Oxidative Biocatalysts

Due to the ubiquitous presence of oxygen and its importance in bioprocesses, screening of biocatalysts in terms of oxygen utilization and/or influence is highly relevant. Of the different enzymes (oxidases, peroxidases and oxygenases) that use oxygen as the electron acceptor during the reaction, the screening of oxygenase variants or mutants is specially complicated. This stems from the fact that the generated products do not cause a change in pH, color or fluorescence, and distinction between regioisomers is often the main objective of the screening program. Also, since most oxygenase substrates and products present poor water solubility, biotransformations are mostly performed in two-liquid phase systems. Furthermore, it is usually applied in whole cell systems and strains that express oxygenases require oxygen not only for the bioconversion but also for endogenous respiration, and so during the biotransformation the oxygen pressure needs to be maintained in order to allow the oxidation reaction to compete with respiration [32]. Among oxygenases, dioxygenases are especially interesting, since they are able to stereo- and regioselectively introduce two oxygen atoms from molecular oxygen in the substrate (e.g., aromatic compounds) [33]. Furthermore, certain types of dioxygenases such as Rieske non-heme iron-dependent oxygenases (ROs), are capable of catalyzing various oxidation reactions (e.g., monohydroxylations, desaturations, oxidative cyclizations) in a variety of substrates, and the range of substrates can be increased (to include non-natural substrates) by changes in the topology of the active site, as demonstrated by Gally et al. (2015) [34].

Screening of dioxygenases is thus usually achieved with more time-consuming and sensitive analytical methods, some of them mentioned above, such as GC or liquid-chromatography (LC) often coupled with MS, or even NMR [32]. These analytical methods can provide a high degree of throughput. However, this throughput is achieved with a lengthy period between variant development, reaction performance and analysis of product concentration and range of compounds. High-throughput screening with robotic microtiter-based devices, as mentioned before, enables the analysis of 100 to 1000 samples of cells per day, but tends to provide hits with low activities and regiospecificity, especially in Gram-positive cells and fungi. Screening can also be performed through enrichment cultures, where the organisms grown on the starting material, might be able to degrade the desired substrate. However, oxidation of the substrate might not occur on the target position [32].

Measurement of the initial oxygen consumption rate in the presence of substrate excess has also been used for characterization of oxygenases, namely dioxygenases [35–37]. A study performed by Parales et al. (1999) used oxygen measurements with Clark-type oxygen electrodes to confirm the importance of an aspartate residue for the catalytic activity in Naphthalene dioxygenase (NDO) enzymes, namely in the electron transfer route. When this residue was modified, no oxygen consumption was observed. Furthermore, in this study a stoichiometric consumption of naphthalene and oxygen for wild-type enzymes was observed [38]. In Rachinskiy et al. (2014) monitoring of oxygen levels was used to detect enzyme deactivation as a parameter in a long-term stability enzyme characterization model to predict the process properties of an enzyme in order to aid enzyme screening for industrial applications [39]. Despite the availability of oxygen sensor spots, as well as other formats of oxygen sensors [40,41], for integration in microtiter plates and shake flasks, the oxygen measurement seems to be used mainly as a monitoring or initial characterization parameter, and not as a screening parameter. Oxygen assay studies performed with purified enzyme solutions have indicated that this could be due to the interference of cell respiration in whole cell solutions (where most of the current screening approaches for this type of enzymes are performed). Moreover, the limited application of oxygen measurement as a screening parameter so far might be related to the lower sensitivity of the traditionally used sensors (Clark-type) or the need of working in closed vessels, when working with highly volatile compounds (and so sensors such as syringe or Clark-type cannot be applied). There are more recently developed, highly sensitive and fast oxygen sensors using optical fibers that can contribute to the increase in oxygen monitoring ability in this field and in turn in the knowledge of the role oxygen plays in these biotransformations (e.g., [40,41]). A system capable of monitoring and measuring oxygen levels and its consumption may provide extra input in screening mutants involved in oxygen dependent reactions. Rate of consumption, oxygen availability in the reaction mixture, diffusion limitation of oxygen and/or other substrates, uncoupling and cell density effects are some of the possible biocatalyst characteristics that can be obtained by using more sensitive and integrated oxygen sensors.

Microfluidic systems can enable a good control of diffusion and gradient generation, due to the predominance of laminar flow and mostly diffusion-limited mass and heat transfer achievable at the micro-scale [42], thus providing the perfect environment for studying the influence of oxygen levels on cells. Therefore, in this work, a microfluidic system along with such type of sensors was applied. A meander microchannel with integrated oxygen sensors [43] was here used to monitor and quantify oxygen consumption rate during the bioconversion of styrene to 1-phenylethanediol by wild-type NDO and two NDO variants in *E. coli*. The *E. coli* cells were used as resting cells, since the use of whole cell biocatalysts as resting cells, besides separating growth phase from catalysis, can moreover decrease the competition of cellular reactions such as oxidative phosphorylation with co-factor regeneration [6]. Styrene bioconversion to 1-phenylethanediol was chosen as the reference reaction since styrene has a similar molecular structure to both the native substrate of the chosen enzymes (naphthalene) and the target substrates of the modified enzymes (different alkenes). Hence, styrene was used to compare the different variants in terms of ability to convert this family of substrates. The chosen case study involved the screening of two previously developed dioxygenase variants [44,45] and their comparison with the wild-type NDO.

In this way, the main goal of this work was to test, as proof-of-concept, whether such a microfluidic system with integrated luminescent oxygen sensors can be used to accelerate the screening of dioxygenase variants, by identifying the earliest reaction time point where a difference in reaction rate can be observed. These reactions are usually performed for 20 h and the mutants evaluated by quantifying product concentration at the end of the reaction by GC [34,44]. By monitoring the oxygen consumption rate at shorter residence times, relative to the conventional approach, during the reaction of different NDO variants, an earlier identification of differences in oxygen consumption may be achieved that may indicate differences in substrate selectivity and/or reaction rate. This may in turn enable a pre-selection of a smaller number of interesting variants to fully test in terms of product identification and quantification with a GC. Thus, the identification of an earlier reaction time where reaction rates are distinct enough to identify a better variant is highly valuable and would furthermore allow a better understanding of the kinetics of the different variants with a potential increase in screening throughput.

2. Materials and Methods

2.1. Materials

All solvents (MTBE, ethanol, 1-octanol), buffer components (potassium phosphate dibasic and monobasic, sodium chloride) and chemicals (styrene, 1-phenylethanediol, indole, IPTG, ampicillin, agarose, yeast extract, tryptone, glycerol, glucose) were obtained from Sigma-Aldrich and Fluka (Steinheim, Germany), Carl Roth GmbH (Karlsruhe, Germany) and Alfa Aesar (Karlsruhe, Germany). *E. coli* JM109/DE3_pDTG141 was obtained by Julia Halder and Prof. Bernhard Hauer (Biocatalysis Group, Institut für Technische Biochemie (ITB), Universität Stuttgart, Stuttgart, Germany) from Prof. Dr. Rebecca Parales (Department of Microbiology and Molecular Genetics, College of Biological Sciences, UC Davis, University of California, Davis, CA, USA) [46].

2.2. Heterologous Expression of Naphthalene Dioxygenase (in E. coli)

The general protocol followed to obtain the variants/mutants is described in Gally et al. (2015) [34] and further optimized towards a better reproducibility as presented in Halder et al. (2017) [45].

To produce induced biomass, *E. coli* JM109 (DE3) cells previously made competent using rubidium chloride, were thawed on ice for 5 min. Then, 1 μL of plasmid DNA for naphthalene dioxygenase (NDO, *Pseudomonas* sp. NCIB 9816-4, pDTG141) or one of the tested mutants was added to the cells and mixed gently by flicking the base of the eppendorf tube and shortly centrifuging. Cell transformation was performed by heat shock by placing the cells in a water bath at 42 °C for 90 s, followed by 2 min on ice. The heat shock treatment was followed by addition of 500 μL of LB medium to the cells and incubation for 1 h at 37 °C and 600 rpm. The competent cells were then plated on selective agar plates containing ampicillin (100 μg/mL) and incubated overnight at 37 °C. To generate the induced cells, one colony from the agar plates was used to inoculate a 2 L shaking flask with 500 mL of TB medium and 500 μL of ampicillin. The flask was incubated at 37 °C and 180 rpm until an optical density (OD_{600nm}) of 0.8–1 was obtained. The cells were then induced with 0.1 mM of isopropyl β-D-1-thiogalactopyranoside (IPTG) dissolved in water and incubated at 25 °C for 16 to 18 h. Indole (Figure 2c) was added to the induced cells as a simple screening test in the solid phase in order to check if induction was achieved, since cells successfully induced with NDO or variants produce indigo, turning the media blue [47–49]. A representation of the molecular structure of the different substrates used and/or mentioned in the text is presented in Figure 2. Influence of indigo (blue color of the cells) in the optical measurement of oxygen was considered negligible since indigo has a wavelength of 420–440 nm while the laser for excitation of the oxygen sensitive dye emits at 620 nm and the detected excitation light from the sensor dye is 760 nm. After induction, the cells were harvested by centrifuging for 20 min at $6000 \times g$ and 4 °C in an Avanti J-26XP centrifuge (from Beckman Coulter, Brea, CA, USA) and resuspended in 0.1 M potassium phosphate buffer (pH 7.2) containing 20 mM glucose.

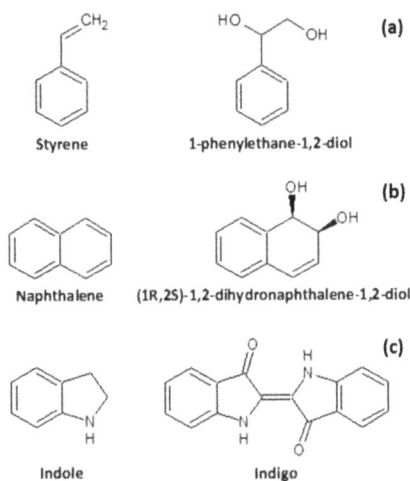

Figure 2. Substrate (left) and corresponding dihydroxylation product (right) for the biotransformation performed (**a**), the natural naphthalene dioxygenase (NDO) substrate (**b**) and the substrate for induction screening, indole (**c**).

2.3. Preparation of Freeze-Dried Cells

To obtain freeze-dried cells, the harvested cells were resuspended in 0.9% sodium chloride (NaCl) solution and centrifuged for 15 min at 4000× *g* and 4 °C. The cells were then spread uniformly in a petri dish and placed inside a freezer at −80 °C for up to 2 h. After freezing the cells were placed inside a freeze drier and freeze-dried overnight under vacuum conditions.

2.4. Biotransformation

The cells for the reaction for gas chromatography (GC) validation were prepared by dissolving 0.1 g_{cww}/mL and 0.05 g_{cww}/mL (cell wet weight) of the freshly prepared resting cells, or 66 mg_{cdw}/mL and 33 mg_{cdw}/mL (cell dry weight) of the freeze-dried resting cells, in 1 mL of 0.1 M phosphate buffer (pH 7.2) with 20 mM glucose. The cells for the oxygen measurements were also dissolved in 1 mL of 0.1 M phosphate buffer (pH 7.2) with 20 mM glucose, but at lower concentrations (0.005 g_{cww}/mL to 0.05 g_{cww}/mL). Glucose was added for in situ co-factor regeneration. Immediately, before starting the reaction, styrene (from a stock solution of 100 mM styrene in pure ethanol) was added to the solution to have 1, 1.5 or 2 mM styrene present for the reaction. The reaction was performed in 4 mL vials with a plastic cap (GC and oxygen measurements) or with a plastic cap with a rubber seal (oxygen measurements) in a tabletop orbital MRH11 Heating ThermoMixer (from HLC BioTech, Bovenden, Germany) at 30 °C and 400 rpm (with 3 mm shaking diameter). The rotation chosen to perform the bioconversion was optimized by previously measuring oxygen concentration with reaction time using an oxygen sensor integrated in a syringe tip (Fixed Oxygen Minisensor OXF500PT from Pyro Science, Aachen, Germany) connected to a FireStingO2 Optical Oxygen Meter (from Pyro Science, Aachen, Germany). The rotation which allowed the reaction to be performed with a constant supply of oxygen was the selected one. Two reaction vials were used per residence time, one for GC validation and one for oxygen measurement.

2.5. GC Analytical Measurement

The samples for GC analysis were prepared by centrifuging the cells and extracting the supernatant two times with 500 µL of methyl tert-butyl ether (MTBE). The reaction mixture was analyzed at different reaction times by measuring product (1-phenylethanediol) (Figure 2a) concentration in the GC/FID-2010 (from Shimadzu, Kyoto, Japan). A Zebron ZB-1 column (30 m × 0.25 mm × 0.25 μm, from Phenomenex, Torrance, CA, USA) was used, with hydrogen as carrier gas (constant pressure of 50.2 kPa) and the injector and detector at 250 °C and 330 °C, respectively. For the detection, the column oven was set at 70 °C for 2 min, then raised to 120 °C at a rate of 15 °C/min and then raised to 320 °C at a rate of 50 °C/min. The gas chromatography with flame ionization detector (GC-FID) was operated in split mode, using 1 mM of 1-octanol in MTBE as standard. The retention times of all the substances measured are 6.084 min for 1-octanol and 7.729 min for 1-phenylethanediol.

2.6. Meander Microfluidic Channel

The used microfluidic channel is a glass and silicon chip developed and batch-produced by iX-factory (now part of Micronit, Enschede, The Netherlands) in Dortmund, Germany, and has been previously described in Ehgartner et al. (2016) [43]. The microchannel has two main inlets and one outlet, with 6 side inlets/outlets. It has a serpentine shape with 18.5 turns, 0.504 m length and a total volume of 10.08 μL, while the main inlet branched-channels have a volume of 0.44 μL. The microchannel is 200 μm deep and 100 μm wide and presents seven chambers with the same geometry of the sensing areas (3.5 mm length and 1 mm width). The final meander microchannel with integrated oxygen sensors is presented in Figure 3.

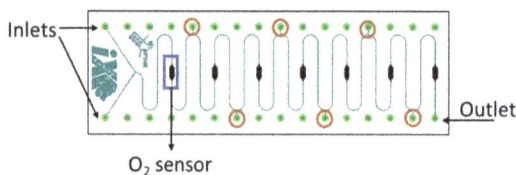

Figure 3. Schematics of the meander microfluidic channel with the main inlets and outlet emphasized. In orange, the five side inlets are highlighted, and in blue one of the seven oxygen sensors is indicated.

2.7. Oxygen Measurement Setup

The setup used for the measurement of the oxygen consumption rate in the cell samples is presented in Figure 4. The reaction of styrene conversion by the different NDOs was performed in a 4 mL vial and sampled at different reaction times inside the vial. The reaction for each of the residence times considered (1, 3, 5, 15 and 30 min) was performed in a different vial. The sampling and subsequent measurement of oxygen consumption rate was performed in the previously described microchannel with integrated oxygen platinum(II) meso-tetra(4-fluorophenyl) tetrabenzoporphyrin (PtTPTBPF) sensors. More details regarding how the oxygen measurement is performed with this sensors can be found in Ehgartner et al. (2016) [43]. As demonstrated in the setup presented in Figure 4, sampling occurs by pulling a certain volume of the reaction mixture from the vial directly inside the microchannel through the outlet of the meander channel, thus allowing a fast measurement of the oxygen consumption in the sample through single-sampling of the reaction mixture.

The setup presented includes two Cavro® XLP 6000 syringe pumps (from Tecan, Männedorf, Switzerland) with 250 μL syringes controlled with LabVIEW (from National Instruments, Austin, TX, USA). The two syringes were connected to the meander microchannel by polytetrafluoroethylene (PTFE) 1.5875 mm (OD) × 1mm (ID) tubing (S 1810-12) (from Bohlender, Grünsfeld, Germany), using Flangeless polypropylene (PP) fingertight 1.5875 mm (ID) fittings (XP-201) and flangeless ferrules (P200X) (from Upchurch Scientific®, Oak Harbor, WA, USA).

The cleaning of the microchannel was performed in between samples with ethanol 5% (*v*/*v*) in deionized water followed by solely deionized water. Deionized water was first pulled through the outlet and then the cleaning solutions were pushed through the channel's second inlet. All the experiments presented in this work were obtained with the meander microfluidic channel placed in an upright position (as presented in Figure 5), to minimize accumulation of cells at the inlet/outlet.

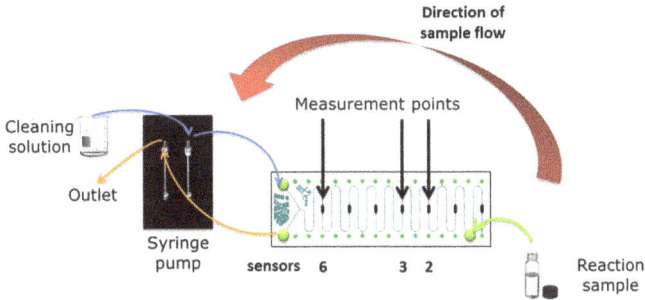

Figure 4. Oxygen measurement setup used for screening variants for styrene biotransformation: pulling approach, with the oxygen sensors position indicated.

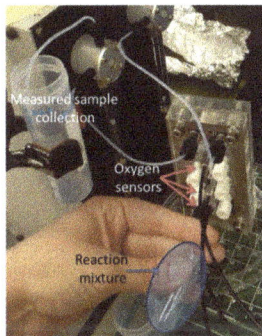

Figure 5. Picture of the pulling sample setup with the meander microchannel in the vertical position.

3. Results

This work aims at exploring the feasibility of using a microreactor with integrated oxygen sensors as an alternative screening technology for dioxygenases. An earlier identification of a difference between two variants would allow an acceleration of the first stage of enzyme screening.

To achieve the proposed goal, an evaluation of the following characteristics was performed:

- The microreactor's ability to distinguish the oxygen consumption due to the reaction from the cells endogenous respiration;
- The microreactor's ability to distinguish the oxygen consumption rate of different dioxygenase variants during the reaction;
- The microreactor's ability to differentiate the oxygen consumption rates during the reaction at various substrate concentrations.

In the beginning of this study, it was thought that the conversion of styrene required 20 h to reach completion. Furthermore, it was unknown at which point of the reaction a difference in oxygen consumption distinct enough to allow enzyme screening might be observed. As residence times of more than 1 h would be difficult to achieve in the used microchannel (due to its dimension), and it was unknown the number of monitoring (residence time) points that would be required to perform the screening, it was deemed best to perform the reaction in a separate vial and only perform the measurement inside the microchannel. As such, per residence time point, 250 μL of the reaction mixture were sampled from the reaction vial and introduced inside the microchannel at the same flowrate. During the uptake, the oxygen concentration inside the channel was measured. The oxygen consumption rate values were determined from the slope of oxygen concentration during the sampling (oxygen concentration divided by measurement time). The variability of the oxygen measurement was quantified during the reaction with empty-vector cells and is presented through the error bars in Figures 6a, 7a, 8a and 9. In Figures 6b, 7b and 8b the error bars represent the reproducibility of the performed GC measurements.

3.1. Reaction vs. Endogenous Respiration

The oxygen consumption rate in the presence of substrate (1 mM of styrene) of empty-vector and wild-type (wt) cells was measured for the selected reaction times: 1, 3, 5, 15 and 30 min, as can be observed in Figure 6a. Less measurements were performed for empty-vector cells since in previous experiments (data not shown) a small change in oxygen consumption value was observed after 3 min, so value at 15 min can be considered as representative of oxygen consumption values of empty-vector cells after 3 min of reaction residence time. Cells containing the wt NDO presented almost 2-fold the oxygen consumption rate of the empty-vector cells in the first 5 min of the reaction. This indicates that there is indeed a higher oxygen consumption in the presence of NDO, which is in principle due to the conversion of styrene by the enzyme. The oxygen consumption rate measured in the three oxygen sensors used (2, 3 and 6) is presented in Figure 6a. Product (1-phenylethanediol) concentration at the different residence times was also measured (Figure 6b) in order to compare with previously obtained results [44,45], and allow a correlation between observed oxygen consumption and substrate conversion. As observed, there is formation of the expected product (1-phenylethanediol) in the presence of the NDO and the high conversion in the initial minutes of the reaction agrees with the observed higher oxygen consumption for the wt cells.

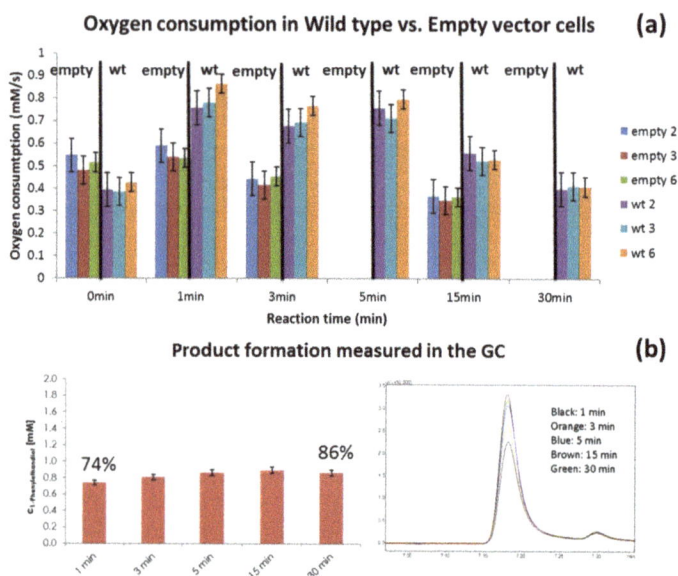

Figure 6. Oxygen measurements of the reaction of empty and wt cells in the presence of 1 mM styrene concentration (**a**); and corresponding GC results for the wt cells of 1-phenylethanediol concentration produced during the reaction and percentage of substrate conversion obtained at 1 and 30 min of residence time (**b**).

3.2. Variant Screening

To evaluate the microreactor's performance in screening dioxygenase variants, the oxygen consumption rate in the presence of styrene (2 mM) of two NDO-variant containing cells (V260A and H295A) was measured in the meander microfluidic channel. Measurements of product concentration at the different reaction times was also performed in the GC.

Figure 7a presents the comparison of the oxygen consumption values obtained for the variants with the ones previously attained for the empty-vector and wt cells. Variant V260A presents a significantly higher oxygen consumption rate, especially in the first 3 min of the reaction, than wt cells and the other variant. When this is considered together with the GC results it seems to indicate that variant V260A has a faster initial reaction rate, achieving also a higher substrate conversion than wt cells and variant H295A. The latter (H295A), on the other hand, presents a relatively stable oxygen consumption rate during the reaction times monitored, which seems to translate into a slower substrate conversion relative to the other NDOs tested (Figure 7b). The product conversion obtained in the last residence time measured does not differ significantly between the different NDOs, but it decreases in the case of V260A, which may indicate the occurrence of a secondary reaction or further oxidation reactions.

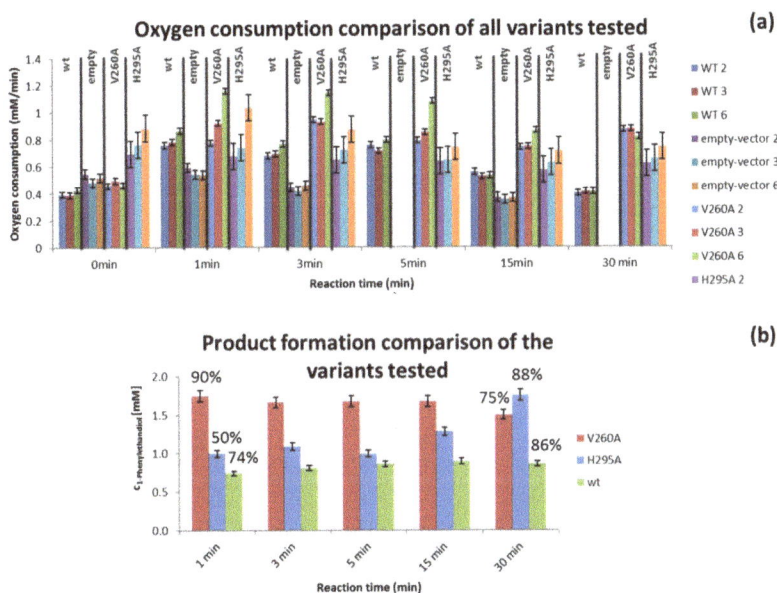

Figure 7. Oxygen consumption rate values for all the cell types tested with the pulling sample approach and higher cell concentration (**a**) and GC results for the two variant containing cells (at 2 mM styrene) and wt cells (at 1 mM) of 1-phenylethanediol concentration produced during the reaction and percentage of substrate conversion obtained at 1 and 30 min of residence time (**b**).

3.3. Substrate Concentration Screening

To test the device's ability to distinguish reaction rate at different substrate concentrations, the oxygen consumption rate and product concentration of wt cells at two different styrene concentrations (1 mM and 1.5 mM) were also measured for the different residence times. No significant difference was observed in terms of oxygen consumption rate, which could be either due to the small difference in substrate concentration or possibly due to mass-transfer limitation of the substrate though the cell wall.

The amount of product obtained was also similar for the two styrene concentrations tested, as can be observed in Figure 8b, which seems to indicate mass-transfer limitation constraints.

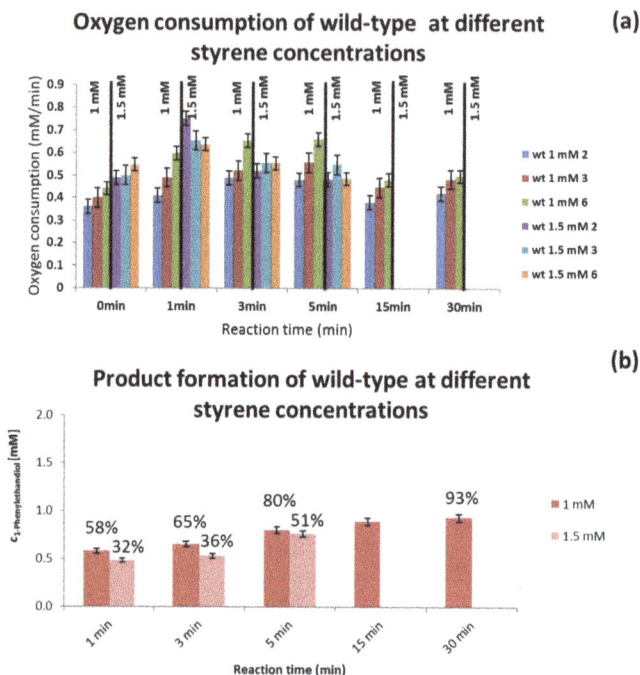

Figure 8. Oxygen (a) and GC (b) measurements of the reaction wt cells in the presence of two styrene concentrations (1 and 1.5 mM).

3.4. Influence of Cell Preparation Method

The influence of cell preparation on cell behavior was assessed by measuring the endogenous respiration of freshly prepared empty-vector cells and freeze-dried empty-vector cells in the presence of 2 mM styrene. As shown in Figure 9, no significant difference in the oxygen consumption rate was observed between the two cell preparations, indicating that it is possible to use both fresh and freeze-dried preparations to perform this screening.

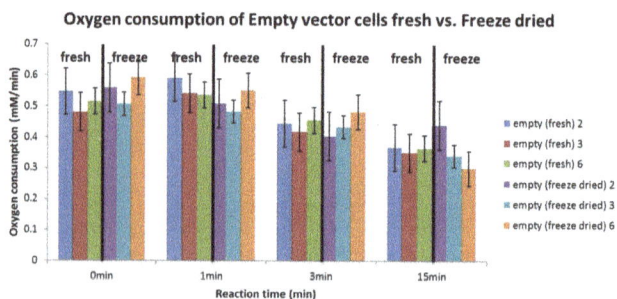

Figure 9. Oxygen consumption rate during the reaction for empty-vector (empty) cells in fresh and freeze-dried preparations at 2 mM styrene.

Bioengineering **2018**, *5*, 30

4. Discussion

The application of the meander channel with integrated oxygen sensors to the measurement of reactions involving whole cells encompassed several challenges. One of the main concerns was the possible clogging of the microchannels due to formation of cell clusters and its effect on fluid flow. *E. coli* cells have dimensions (<2 μm in length) smaller than the microchannels in the meander chip used, and so no significant obstruction of the channels was expected for the relatively diluted solutions used in the experimental work. Another concern involved the cell endogenous respiration rate, which in case of high respiration rates could rapidly consume the oxygen inside the reactor and thus prevent the measurement of the oxygen consumption due to the bioconversion. To address both issues, the cell density was optimized. This was achieved by introducing in the microreactor at a fixed flowrate different dilutions of the wild-type (wt) NDO containing resting cells without substrate present. The resting cells used were *E. coli* JM109 (DE3) cells which were transformed with a pDTG141 plasmid for NDO. The cell dilution considered optimal (0.01 g_{cww}/mL) was the one which allowed a good distinction between slow and fast reaction rates, as shown in the Table S1 in Supplementary Materials. The influence of the substrate (styrene) on the oxygen sensors used (which contain polystyrene), was also considered and some tests performed, which are described in the Supplementary Materials.

4.1. Measurement Setup

Ideally, the meander microchannel with integrated oxygen sensors would function as an online monitoring system, where the inlet tube connected to the syringe pump would be submerged in the reaction volume, thus allowing sampling from the reaction mixture at given reaction times. This approach, however, was not followed in this study due to several difficulties that could arise from such setup. One of these issues is regarding the high volatility of the substrate (styrene), which could lead to a decrease in the concentration over time, since to sample the reaction mixture, the vial would not be completely sealed. The use of an adequate membrane or a septum could avoid this issue. The other concern relates to the fact that the reaction could be faster than expected and the initial reaction rate missed due to the time taken for sample uptake into the syringe and then introduction in the microreactor.

To avoid these possible issues, a different sampling strategy was applied. The reaction was performed in two identical vials per residence time placed at the same conditions in the tabletop thermomixer. One of the vials was used for measuring the oxygen consumption rate and the other for product quantification in the GC. This enabled a direct correlation between the oxygen measurements and product concentration at the chosen reaction times. Furthermore, by performing the reaction this way, loss of substrate due to its volatility was minimized since the vial was only opened immediately before measurement. To address the possibility of a fast initial reaction rate, sampling uptake was performed through the outlet, instead of pushing the sample through the inlet (see Figure 5). Through this process, by having a short tube at the outlet the sampling time was just a few seconds, instead of close to 1 min, thus allowing monitoring of the reaction at short reaction times (1, 3 and 5 min) and to observe fast initial reaction rates.

4.2. Reaction vs. Endogenous Respiration

The chosen sampling strategy allowed measurement of a difference in the oxygen consumption rate between empty and wt cells, as can be observed in Figure 6. Cells containing the wt NDO presented almost 2-fold the oxygen consumption rate of the empty-vector cells.

It is interesting to observe that 74% of the substrate is converted in the first minute of the reaction. The product concentrations obtained were equivalent to previous lab-scale assays where the reaction had run for 20 h with only end-point GC quantification [45]. After 15 min of bioconversion the oxygen consumption rate decreased, and a stabilization of product concentration is observed, which seems to fit the oxygen consumption rate profile. The reaction was therefore significantly faster than expected and could be monitored with sampling points until 30 min of reaction time was reached. This allows a significant increase in the number of variants and reactions that can be screened per day with the standard method. It is also a good example of the valuable input that microfluidic systems can provide in terms of reaction kinetics in this field.

4.3. Variant Screening

The oxygen consumption rate of the two NDO-variant containing cells (V260A and H295A) was also measured in the meander microchannel. As can be observed in Figure 7a, a difference between the two variants could be observed, as well as differences in oxygen consumption rate relatively to the wt cells and the background (empty cells). One of the variants (V260A) presented a higher oxygen consumption rate, especially in the first 3 min of the reaction. Since we cannot observe what happens in the reaction in the first minute, by being able to observe a higher oxygen consumption, relative to the previous measurements with wt cells, it could indicate that this reaction had a slower initial reaction rate. Thus, the observation of the time at which the oxygen consumption is fastest was possible after the first minute of the reaction. Nonetheless, it might also simply mean that this variant has a higher reaction rate, leading to a faster oxygen consumption rate. As can be observed in Figure 7b, the latter was verified, since a 90% conversion of the substrate was measured in the first minute of the reaction. This variant however, presents a decrease in product concentration at 30 min (Figure 7b) which may indicate further oxidation of the product into another compound or even the production of the other enantiomer of 1-phenylethanediol.

The other variant (H295A) presented an oxygen consumption rate closer to the wt cells, meaning that the oxygen consumption rate during the bioconversion is very similar to the endogenous respiration for this variant (value of oxygen consumption rate at 0 min in Figure 7a). The observed oxygen consumption rate is also maintained longer above the value for empty cells than for the wt cells. As can be observed in Figure 7, it is also the variant (H295A) that presents the lowest initial conversion with a steady increase during the 30 min of reaction monitored, while the other two cell types maintain approximately the same value after 3 min. Although H295A has a smaller reaction rate than the other measured enzymes, the increased cell respiration might indicate an interference of the heterologous enzyme with the metabolism of the host organism, which is worth of further investigation.

From the presented experiments for empty-vector cells, wt, V260A and H295A, it is possible to conclude that higher consumption rates indicate higher substrate conversion. Moreover, the duration of the oxygen consumption rate above empty-vector cell endogenous respiration levels represents a continuation of substrate conversion, while the decrease in oxygen levels translates in a conservation of product concentration without significant reaction rate.

4.4. Substrate Concentration Screening

To understand the applicability of the meander channel as a screening platform, and whether it is possible to distinguish the oxygen consumption rates due to different substrate concentrations, further experiments were performed. The bioconversion with wt cells was performed at two different concentrations of styrene. As demonstrated in Figure 8a, a slightly higher oxygen consumption rate was observed for the higher styrene concentration, but only in the first minute of reaction. The results of substrate conversion obtained from the GC (Figure 8b) indicate that higher substrate concentration results in a lower initial substrate conversion. This can imply some effect of the substrate on the reaction rate. Nevertheless, since the measured product concentrations for the two reactions are quite similar, this might indicate that the reaction rate is limited by substrate diffusion through the cell wall at the concentrations tested. This mass transport limitations are a well-known issue when using whole cell biosystems, which can be improved by permeabilization of the wall with chemical (e.g., adding organic solvents, surfactants, chelating agents or altering the cell wall's fatty acid content) or physical (e.g., temperature shock, electroporation) methods [50]. Other approaches involve the expression of membrane transporters, to increase influx of substrate to the cell, or the use of cell surface techniques to display the enzymes on the cell membrane, and even temperature-controlled pore-formation through the use of lytic phage proteins [51].

Since no differences between styrene concentration were observed, the distinct oxygen consumption rates observed for the wt cells and the two variants, arise solely from differences in the mutated NDOs.

4.5. Influence of Cell Preparation Method

A further test was performed, comparing the measurement performance of freshly prepared cells and freeze-dried cells, since the cell preparation procedure is considerably longer than the time required for the oxygen measurement (Figure 9). The use of freeze-dried cells would enable to perform tests on the same cells for more than one day while guaranteeing their stability and comparative behavior, as well as analyze older cell samples or samples from different sources (e.g., other laboratory facilities) if required. As can be observed in Figure 9, the oxygen consumption rate for wt and empty cells is still very similar for both cell preparations, and so measurements with freeze-dried cells can not only be performed but also compared with the ones made on freshly prepared cells.

4.6. Evaluation of Oxygen Measurement in the Microchannel as a Screening Method

To the best of our knowledge, the oxygen consumption rate for dioxygenases in whole cells as presented here has not been previously investigated. Hence, the attained values during the experiments presented here were compared with the only values found for pure dioxygenases (measured with the polarographic method) [35]. The comparison between the values is presented below in Table 1. The literature values (in blue in Table 1) cited in the table were calculated from the values presented in the articles, considering the enzyme concentration (an average value of 250 kDa for the oxygenase component was used for the calculations) used in the respective assays, to have comparable units. The values of this study (in gray in Table 1), are composed by an average of the rates obtained in the first three minutes of the reaction (when the highest rates are measured) in sensor 6 (where the more defined oxygen behavior seems to be observed), calculated without the background value of cell respiration.

Table 1. Comparison of oxygen consumption rate/oxygen uptake of the whole cell catalyst (in grey) with values for pure dioxygenases found in literature (in blue) [36,37].

Biocatalyst	Substrate	Amount of Oxygenase Component of NDO	Measurement Setup	Oxygen Uptake (mM·min^{-1})
NDO from *Sphingomonas* CHY-1 [37]	Naphthalene (0.1 mM)	PhnI (0.13 µM)	Clark-type oxygen electrode	14.95
NDO from *Pseudomonas* sp. Strain NCIB 9816-4 [36]	Styrene (0.1 mM)	ISP$_{NAP}$ (25 µg)		0.0700
NDO (pDTG141) wild-type in *E. coli* JM109/DE3)	Styrene (1 mM)			0.3202
NDO variant V260A in *E. coli* JM109/DE3)	Styrene (2 mM)		Luminescent oxygen sensors	0.6553
NDO variant H295A in *E. coli* JM109/DE3)	Styrene (2 mM)			0.4560

NDO: Naphthalene dioxygenase.

A direct comparison between the values retrieved from the literature and the ones measured in this study is difficult to perform since different enzymes, substrates, substrate concentrations and oxygen measurement techniques were used. There are one or two orders of magnitude of difference in the oxygen uptake between the different catalysts, but it is interesting to notice that the highest uptake rate was obtained for the natural substrate of the enzyme (naphthalene). When comparing biocatalyst performance for the same substrate (styrene), it should be noted that the measurement with whole cells was performed at a substrate concentration that is 10-fold higher relative to the corresponding substrate literature value. If a linear relationship between oxygen concentration rate and substrate concentration is considered, then the rate values obtained for the NDO from *E. coli* JM109/DE3 are half or lower than the one obtained for NDO from *Pseudomonas* sp. NCIB 9816-4. The lower oxygen consumption rate could be related to the already discussed diffusion limitations due to the cell membrane. It should, however, be taken into account that the data presented in this work indicates that the oxygen consumption rates measured for the NDO from *E. coli* JM109/DE3 are diffusion limited and so a linear relationship between oxygen concentration rate and substrate concentration does not occur.

It is also worth mentioning the degree of variability in the data presented here, especially in oxygen consumption rate results which was not possible to quantify appropriately throughout the duration of the experiments. Data variability is most likely related to differences in the preparations of the cell solutions due to small variations in the amount of weighed cells, and thus the number of cells present. However, the variability in oxygen data obtained can also be inherent to the measurement itself, as also described by Jouanneau et al. (2006) with the observed discrepancies in the enzyme activities obtained when the oxygen assay was applied (polarographic assay using Clark-type oxygen sensors) [37].

The presented microfluidic system cannot compete in terms of throughput with most of the screening systems discussed in the introduction. However, it can provide a different type of input (oxygen consumption rate and maybe other oxidative properties) than the above discussed screening platforms. In terms of throughput, the presented system can perform a single measurement every 10 min (including sample uptake and channel cleaning). This means that in a continuous operation, it could perform 47 single measurements in an 8 h working day and 129 single measurements in a 22 h working day (with 2 h for thorough cleaning of the microreactor, tubing and system components). Considering the same reactions performed for this work (around 50 min for each one to be completed), per day we could perform 9 entire reactions in an 8 h working day or 26 entire reactions in a 22 h working day. Thus, to perform measurements every minute of a certain reaction vial, 10 microreactors of this type would have to work in parallel, and such a system is easily achieved with microfluidic units. Furthermore, by working in solution and not in droplet, such a system enables a direct measurement (e.g., side-loop with recirculation) from the reaction vial using a simpler microfluidic arrangement, if a more air-tight setup is assembled. This setup could be achieved by using stainless steel or a less permeable polymer as the tubing material, as well as more appropriate sealed reaction vial caps. The system can also be connected to one of the droplet-based systems discussed in the introduction to achieve a more comprehensive characterization of the mutants during screening.

Moreover, two different screening approaches can be achieved with this platform: variant comparison or substrate comparison. The first approach, which was the one demonstrated in this study, compares the oxygen consumption of different variants relative to the same substrate. A higher rate indicates either higher variant activity or higher substrate specificity for the target substrate. In the second approach, the same cell type can be tested with different substrates. As mentioned, different rates would correlate with differences in specificity and activity. Furthermore, in this case, endogenous cell respiration of the cell without the target enzyme, or of the mutant at different substrate concentrations could further indicate possible toxicity concerns from the substrates being screened.

It is also relevant to mention that the bioconversion at the reaction conditions and cell densities used has a relatively high conversion rate (50 to 70% depending on initial substrate concentration), which greatly contributed to the measurable oxygen consumption difference between the biotransformation and cell respiration. Since higher cell densities might lead to obstruction issues, there is a limitation on the use of this platform in terms of reaction rate to fast reactions only, which generate oxygen consumption rates higher than the host's respiration rate. This limitation can be observed in Figure 7 for the H295A variant. Reactions with slower kinetics, which is usually the case for initial dioxygenase variants, might hence be more difficult to detect with this microfluidic platform. Faster reaction rates, on the other hand, can be more easily screened, by increasing the flowrate used during the detection and/or by decreasing the cell density used for the biotransformation or for the measurement.

5. Conclusions

The presented results have demonstrated the potential of the meander microchannel with integrated oxygen sensors to function as a biocatalyst screening platform for oxygen-dependent whole cell biocatalysts. It was possible to measure the endogenous respiration rate of the *E. coli* cells used and distinguish respiration rate from oxygen consumption due to styrene oxidation by wild-type NDO. Additionally, the two tested NDO variants showed different oxygen consumption rates from the ones obtained with the wild-type. Variant V260A presented a higher oxygen consumption rate in the first 3 min of the reaction, corresponding to a faster styrene conversion (90% after 1 min). On the other hand, variant H295A showed a similar oxygen consumption rate compared to the wild-type NDO, which was equally maintained throughout the monitoring time (30 min), unlike the wild-type NDO, whose consumption rate decreased close to endogenous respiration levels after 15 min. H295A variant, generated however a smaller conversion rate with a steady increase during the reaction time, indicating a slower reaction rate than the other variants tested. Furthermore, this variant (H295A) allowed highlighting of a potential limitation in terms of reaction rates measurable by the microfluidic platform, identifying the reaction rate below which the microfluidic platform is unable to distinguish oxygen consumption due to the biotransformation from the endogenous respiration. The oxygen consumption rates measured with the meander channel with integrated oxygen sensors appear to be concordant with the values found in the literature.

The presented platform shows a considerable potential as a screening platform for mutants involved in oxidative reactions, especially enzymes with high reaction rates and fast bioconversions. The platform can achieve a throughput (129 single measurements or 26 reactions per day) comparable with the current approach, which can be further improved by adding more microfluidic systems in parallel and automating sampling. The current throughput of the platform is similar to the one obtained with the GC, where 129 reactions could be measured within 21.5 h, without time required for reaction preparation (which would imply a further 6 h).

In the future, towards achieving a full characterization of the presented microfluidic system as a screening platform, it would be interesting to perform experiments with additional variants to detect relevant patterns or tendencies observable by using only the oxygen sensors. It would also be interesting to check whether phenomena such as product overoxidation beyond the desired product [32], substrate inhibition or toxicity, or uncoupling effects can be observed or detected through oxygen consumption rate measurements. Uncoupling effects can lead to an increase in oxygen demand, as well as the production of toxic hydrogen peroxide and a lowered specific activity in the final bioconversion. This effect can occur in the absence of substrate, when the substrate cannot be oxidized or in the presence of compounds that do not properly fit the active site [32]. Furthermore, since it was observed that the styrene conversion occurs much faster than initially thought, in a future iteration of the work presented here, the use of the highly controlled flow and residence time dependencies characteristic of microfluidic systems will be used to further characterize the conversion reaction. In this case, the whole cell catalyst will be introduced through one of the channel inlets, and the substrate (or different substrates) introduced through the other. The introduction of the different components inside the microchannel may allow the collection of initial rate values, and a more detailed understanding of the first minute of the reaction.

Supplementary Materials: The following are available online at http://www.mdpi.com/2306-5354/5/2/30/s1. The cell concentration used for the experiments is highlighted in green.

Acknowledgments: The authors thank the funding from the People Programme (Marie Curie Actions, Multi-ITN) of the European Union's Seventh Framework Programme for research, technological development, and demonstration under grant agreement no 608104 (EuroMBR).

Conflicts of Interest: The authors declare no conflict of interest.

Abbreviations

NDO	Naphthalene dioxygenase
E. coli	*Escherichia coli*
GC	Gas chromatography
GC-FID	Gas chromatography with flame ionization detector
GFP	Green fluorescent protein
FACS	Fluorescence-activated cell sorting
MS	Mass spectrometry
HPLC	High-performance liquid chromatography
LC	Liquid-chromatography
DESI	Desorption electrospray ionization
NMR	Nuclear magnetic resonance
NIR	Near infrared
ROs	Rieske non-heme iron-dependent oxygenases
PTFE	Polytetrafluoroethylene
wt	Wild-type cells
empty	Empty-vector cells
pDTG141	The NDO-containing plasmid used in this work

References

1. Adrio, J.L.; Demain, A.L. Microbial enzymes: Tools for biotechnological processes. *Biomolecules* **2014**, *4*, 117–139. [CrossRef] [PubMed]
2. Burton, S.G.; Cowan, D.A.; Woodley, J.M. The search for the ideal biocatalyst. *Nat. Biotechnol.* **2002**, *20*, 37–45. [CrossRef] [PubMed]
3. Tufvesson, P.; Lima-Ramos, J.; Nordblad, M.; Woodley, J.M. Guidelines and cost analysis for catalyst production in biocatalytic processes. *Org. Process Res. Dev.* **2011**, *15*, 266–274. [CrossRef]
4. Ishige, T.; Honda, K.; Shimizu, S. Whole organism biocatalysis. *Curr. Opin. Chem. Biol.* **2005**, *9*, 174–180. [CrossRef] [PubMed]
5. Ladkau, N.; Schmid, A.; Bühler, B. The microbial cell-functional unit for energy dependent multistep biocatalysis. *Curr. Opin. Biotechnol.* **2014**, *30*, 178–189. [CrossRef] [PubMed]
6. Kuhn, D.; Blank, L.M.; Schmid, A.; Bühler, B. Systems biotechnology—Rational whole-cell biocatalyst and bioprocess design. *Eng. Life Sci.* **2010**, *10*, 384–397. [CrossRef]
7. Ogawa, J.; Shimizu, S. Microbial enzymes: New industrial applications from traditional screening methods. *Trends Biotechnol.* **1999**, *17*, 13–21. [CrossRef]
8. Packer, M.S.; Liu, D.R. Methods for the directed evolution of proteins. *Nat. Rev. Genet.* **2015**, *16*, 379–394. [CrossRef] [PubMed]
9. Doig, S.D.; Pickering, S.C.R.; Lye, G.J.; Woodley, J.M. The use of microscale processing technologies for quantification of biocatalytic Baeyer-Villiger oxidation kinetics. *Biotechnol. Bioeng.* **2002**, *80*, 42–49. [CrossRef] [PubMed]
10. Van Rossum, T.; Kengen, S.W.M.; Van Der Oost, J. Reporter-based screening and selection of enzymes. *FEBS J.* **2013**, *280*, 2979–2996. [CrossRef] [PubMed]
11. Reetz, M.T.; Becker, M.H. A Method for High-Throughput Screening of Enantioselective Catalysts. *Angew. Chem. Int. Ed.* **1999**, *38*, 1758–1761. [CrossRef]
12. Yan, C.; Parmeggiani, F.; Jones, E.A.; Claude, E.; Hussain, S.A.; Turner, N.J.; Flitsch, S.L.; Barran, P.E. Real-time screening of biocatalysts in live bacterial colonies. *J. Am. Chem. Soc.* **2017**, *139*, 1408–1411. [CrossRef] [PubMed]
13. Heuts, D.P.H.M.; Van Hellemond, E.W.; Janssen, D.B.; Fraaije, M.W. Discovery, characterization, and kinetic analysis of an alditol oxidase from Streptomyces coelicolor. *J. Biol. Chem.* **2007**, *282*, 20283–20291. [CrossRef] [PubMed]
14. Quake, S.R.; Fu, A.Y.; Spence, C.; Scherer, A.; Arnold, F.H. A microfabricated fluorescence-activated cell sorter. *Nat. Biotechnol.* **1999**, *17*, 1109–1111. [CrossRef]

15. Fernández-Álvaro, E.; Snajdrova, R.; Jochens, H.; Davids, T.; Böttcher, D.; Bornscheuer, U.T. A combination of in vivo selection and cell sorting for the identification of enantioselective biocatalysts. *Angew. Chem. Int. Ed.* **2011**, *50*, 8584–8587. [CrossRef] [PubMed]

16. Bornscheuer, U.T.; Huisman, G.W.; Kazlauskas, R.J.; Lutz, S.; Moore, J.C.; Robins, K. Engineering the third wave of biocatalysis. *Nature* **2012**, *485*, 185–194. [CrossRef] [PubMed]

17. Parmeggiani, F.; Lovelock, S.L.; Weise, N.J.; Ahmed, S.T.; Turner, N.J. Synthesis of D- and L-Phenylalanine Derivatives by Phenylalanine Ammonia Lyases: A Multienzymatic Cascade Process. *Angew. Chem. Int. Ed.* **2015**, *54*, 4608–4611. [CrossRef] [PubMed]

18. Joo, H.; Arisawa, A.; Lin, Z.; Arnold, F.H. A high-throughput digital imaging screen for the discovery and directed evolution of oxygenases. *Chem. Biol.* **1999**, *6*, 699–706. [CrossRef]

19. Schwaneberg, U.; Schmidt-Dannert, C.; Schmitt, J.; Schmid, R.D. A Continuous Spectrophotometric Assay for P450 BM-3, a Fatty Acid Hydroxylating Enzyme, and Its Mutant F87A. *Anal. Biochem.* **1999**, *269*, 359–366. [CrossRef] [PubMed]

20. Duetz, W.A.; Rüedi, L.; Hermann, R.; O'Connor, K.; Büchs, J.; Witholt, B. Methods for intense aeration, growth, storage, and replication of bacterial strains in microtiter plates. *Appl. Environ. Microbiol.* **2000**, *66*, 2641–2646. [CrossRef] [PubMed]

21. Schwaneberg, U.; Otey, C.; Cirino, P.C.; Farinas, E.; Arnold, F. Cost-Effective Whole-Cell Assay for Laboratory Evolution of Hydroxylases in *Escherichia coli*. *J. Biomol. Screen.* **2001**, *6*, 111–117. [CrossRef] [PubMed]

22. Samorski, M.; Müller-Newen, G.; Büchs, J. Quasi-continuous combined scattered light and fluorescence measurements: A novel measurement technique for shaken microtiter plates. *Biotechnol. Bioeng.* **2005**, *92*, 61–68. [CrossRef] [PubMed]

23. Clouthier, C.M.; Pelletier, J.N. Expanding the organic toolbox: A guide to integrating biocatalysis in synthesis. *Chem. Soc. Rev.* **2012**, *41*, 1585. [CrossRef] [PubMed]

24. Viefhues, M.; Sun, S.; Valikhani, D.; Nidetzky, B.; Vrouwe, E.X.; Mayr, T.; Bolivar, J.M. Tailor-made resealable micro(bio)reactors providing easy integration of in situ sensors. *J. Micromech. Microeng.* **2017**, *27*. [CrossRef]

25. Couniot, N.; Francis, L.A.; Flandre, D. A 16 × 16 CMOS Capacitive Biosensor Array Towards Detection of Single Bacterial Cell. *IEEE Trans. Biomed. Circuits Syst.* **2016**, *10*, 364–374. [CrossRef] [PubMed]

26. Im, H.; Shao, H.; Park, Y., Il; Peterson, V.M.; Castro, C.M.; Weissleder, R.; Lee, H. Label-free detection and molecular profiling of exosomes with a nano-plasmonic sensor. *Nat. Biotechnol.* **2014**, 490–495. [CrossRef] [PubMed]

27. Wohlgemuth, R.; Plazl, I.; Žnidaršič-Plazl, P.; Gernaey, K.V.; Woodley, J.M. Microscale technology and biocatalytic processes: Opportunities and challenges for synthesis. *Trends Biotechnol.* **2015**, *33*, 302–314. [CrossRef] [PubMed]

28. Abate, A.R.; Ahn, K.; Rowat, A.C.; Baret, C.; Marquez, M.; Klibanov, A.M.; Grif, A.D.; Weitz, D.A.; Aga, G.A.L. Ultrahigh-throughput screening in drop-based microfluidics for directed evolution. *Proc. Natl. Acad. Sci. USA* **2010**, *107*, 6550–6550. [CrossRef]

29. Kintses, B.; Hein, C.; Mohamed, M.F.; Fischlechner, M.; Courtois, F.; Lainé, C.; Hollfelder, F. Picoliter cell lysate assays in microfluidic droplet compartments for directed enzyme evolution. *Chem. Biol.* **2012**, *19*, 1001–1009. [CrossRef] [PubMed]

30. Horka, M.; Sun, S.; Ruszczak, A.; Garstecki, P.; Mayr, T. Lifetime of Phosphorescence from Nanoparticles Yields Accurate Measurement of Concentration of Oxygen in Microdroplets, Allowing One To Monitor the Metabolism of Bacteria. *Anal. Chem.* **2016**, *88*, 12006–12012. [CrossRef] [PubMed]

31. Hess, D.; Rane, A.; Demello, A.J.; Stavrakis, S. High-throughput, quantitative enzyme kinetic analysis in microdroplets using stroboscopic epifluorescence imaging. *Anal. Chem.* **2015**, *87*, 4965–4972. [CrossRef] [PubMed]

32. Van Beilen, J.B.; Duetz, W.A.; Schmid, A.; Witholt, B. Practical issues in the application of oxygenases. *Trends Biotechnol.* **2003**, *21*, 170–177. [CrossRef]

33. Cirino, P.C.; Arnold, F.H. Protein engineering of oxygenases for biocatalysis. *Curr. Opin. Chem. Biol.* **2002**, *6*, 130–135. [CrossRef]

34. Gally, C.; Nestl, B.M.; Hauer, B. Engineering Rieske Non-Heme Iron Oxygenases for the Asymmetric Dihydroxylation of Alkenes. *Angew. Chem. Int. Ed.* **2015**, *54*, 12952–12956. [CrossRef] [PubMed]

35. Whittaker, J.W.; Orville, A.M.; Lipscomb, J.D. [14] Protocatechuate 3,4-dioxygenase from *Brevibacterium fuscum*. *Methods Enzymol.* **1990**, *188*, 82–88. [CrossRef] [PubMed]

36. Lee, K.; Gibson, D.T. Stereospecific dihydroxylation of the styrene vinyl group by purified naphthalene dioxygenase from *Pseudomonas* sp. strain NCIB 9816-4. *J. Bacteriol.* **1996**, *178*, 3353–3356. [CrossRef] [PubMed]

37. Jouanneau, Y.; Meyer, C.; Jakoncic, J.; Stojanoff, V.; Gaillard, J. Characterization of a naphthalene dioxygenase endowed with an exceptionally broad substrate specificity toward polycyclic aromatic hydrocarbons. *Biochemistry* **2006**, *45*, 12380–12391. [CrossRef] [PubMed]

38. Parales, R.E.; Parales, J.V.; Gibson, D.T. Aspartate 205 in the catalytic domain of naphthalene dioxygenase is essential for activity. *J. Bacteriol.* **1999**, *181*, 1831–1837. [PubMed]

39. Rachinskiy, K.; Kunze, M.; Graf, C.; Schultze, H.; Boy, M.; Büchs, J. Extension and application of the "enzyme test bench" for oxygen consuming enzyme reactions. *Biotechnol. Bioeng.* **2014**, *111*, 244–253. [CrossRef] [PubMed]

40. Sun, S.; Ungerböck, B.; Mayr, T. Imaging of oxygen in microreactors and microfluidic systems. *Methods Appl. Fluoresc.* **2015**, *3*, 34002. [CrossRef] [PubMed]

41. Gruber, P.; Marques, M.P.C.; Szita, N.; Mayr, T. Integration and application of optical chemical sensors in microbioreactors. *Lab Chip* **2017**. [CrossRef] [PubMed]

42. Chiu, D.T.; de Mello, A.J.; Di Carlo, D.; Doyle, P.S.; Hansen, C.; Maceiczyk, R.M.; Wootton, R.C.R. Small but Perfectly Formed? Successes, Challenges, and Opportunities for Microfluidics in the Chemical and Biological Sciences. *Chem* **2017**, *2*, 201–223. [CrossRef]

43. Ehgartner, J.; Sulzer, P.; Burger, T.; Kasjanow, A.; Bouwes, D.; Krühne, U.; Klimant, I.; Mayr, T. Online analysis of oxygen inside silicon-glass microreactors with integrated optical sensors. *Sens. Actuators B Chem.* **2016**, *228*, 748–757. [CrossRef]

44. Halder, J.M. Naphthalene Dioxygenase from Pseudomonas sp. NCIB 9816-4: Systematic Analysis of the Active Site. Ph.D. Thesis, Stuttgart Universität, Stuttgart, Germany, October 2017.

45. Halder, J.M.; Nestl, B.M.; Hauer, B. Semirational Engineering of the Naphthalene Dioxygenase from *Pseudomonas* sp. NCIB 9816-4 towards Selective Asymmetric Dihydroxylation. *ChemCatChem* **2018**, *10*, 178–182. [CrossRef]

46. Parales, R.E.; Resnick, S.M.; Yu, C.L.; Boyd, D.R.; Sharma, N.D.; Gibson, D.T. Regioselectivity and enantioselectivity of naphthalene dioxygenase during arene cis-dihydroxylation: Control by Phenylalanine 352 in the alfa subunit. *J. Bacteriol.* **2000**, *182*, 5495–5504. [CrossRef] [PubMed]

47. Ensley, B.D.; Ratzkin, B.J.; Osslund, T.D.; Simon, M.J.; Wackett, L.P.; Gibson, D.T. Expression of Naphthalene Oxidation Genes in *Escherichia coli* Results in the Biosynthesis of Indigo. *Science* **1983**, *222*, 167–169. [CrossRef] [PubMed]

48. Berry, A.; Dodge, T.C.; Pepsin, M.; Weyler, W. Application of metabolic engineering to improve both the production and use of biotech indigo. *J. Ind. Microbiol. Biotechnol.* **2002**, *28*, 127–133. [CrossRef] [PubMed]

49. Pathak, H.; Madamwar, D. Biosynthesis of indigo dye by newly isolated naphthalene-degrading strain *Pseudomonas* sp. HOB1 and its application in dyeing cotton fabric. *Appl. Biochem. Biotechnol.* **2010**, *160*, 1616–1626. [CrossRef] [PubMed]

50. De Carvalho, C.C.C.R. Enzymatic and whole cell catalysis: Finding new strategies for old processes. *Biotechnol. Adv.* **2011**, *29*, 75–83. [CrossRef] [PubMed]

51. Wachtmeister, J.; Rother, D. Recent advances in whole cell biocatalysis techniques bridging from investigative to industrial scale. *Curr. Opin. Biotechnol.* **2016**, *42*, 169–177. [CrossRef] [PubMed]

bioengineering

MDPI

Article

Metabolic Reprogramming and the Recovery of Physiological Functionality in 3D Cultures in Micro-Bioreactors [†]

Krzysztof Wrzesinski [1,2] and Stephen J. Fey [1,2,*]

[1]   Tissue Culture Engineering Laboratory, Department of Biochemistry and Molecular Biology,
      University of Southern Denmark, 5230 Odense, Denmark; kwr@celvivo.com
[2]   CelVivo IVS, 5491 Blommenslyst, Denmark
*   Correspondence: sjf@celvivo.com; Tel.: +45-5117-7227
†   This article is dedicated to the memory of Vasco Botelho Carvalho.

Received: 22 January 2018; Accepted: 24 February 2018; Published: 7 March 2018

Abstract: The recovery of physiological functionality, which is commonly seen in tissue mimetic three-dimensional (3D) cellular aggregates (organoids, spheroids, acini, etc.), has been observed in cells of many origins (primary tissues, embryonic stem cells (ESCs), induced pluripotent stem cells (iPSCs), and immortal cell lines). This plurality and plasticity suggest that probably several basic principles promote this recovery process. The aim of this study was to identify these basic principles and describe how they are regulated so that they can be taken in consideration when micro-bioreactors are designed. Here, we provide evidence that one of these basic principles is hypoxia, which is a natural consequence of multicellular structures grown in microgravity cultures. Hypoxia drives a partial metabolic reprogramming to aerobic glycolysis and an increased anabolic synthesis. A second principle is the activation of cytoplasmic glutaminolysis for lipogenesis. Glutaminolysis is activated in the presence of hypo- or normo-glycaemic conditions and in turn is geared to the hexosamine pathway. The reducing power needed is produced in the pentose phosphate pathway, a prime function of glucose metabolism. Cytoskeletal reconstruction, histone modification, and the recovery of the physiological phenotype can all be traced to adaptive changes in the underlying cellular metabolism. These changes are coordinated by mTOR/Akt, p53 and non-canonical Wnt signaling pathways, while myc and NF-kB appear to be relatively inactive. Partial metabolic reprogramming to aerobic glycolysis, originally described by Warburg, is independent of the cell's rate of proliferation, but is interwoven with the cells abilities to execute advanced functionality needed for replicating the tissues physiological performance.

Keywords: bioreactors; 3D cell culture; spheroids; organoids; hypoxia; aerobic glycolysis; glutaminolysis; metabolic reprogramming; physiological performance; Warburg

---

1. Introduction

Three-dimensional (3D) cell culture offers a glimpse into tissue function that is only recently being appreciated. This is true whether primary, immortal or stem cells are used and it appears to apply to all types of tissue.

What drives this process? Often the growth conditions used for these two-dimensional (2D) or 3D cell culture studies are identical: the same growth media, temperature, and atmosphere. Often the same cells are used and yet the performance of the cells is radically different. Here we examine the role of metabolic reprogramming and factors that induce the recovery or development of mimetic tissues.

Primary cells retain their physiological behaviour longer when grown in 3D culture conditions (e.g., astrocytes [1], prostate [2], and microvascular networks [3]). Numerous different colorectal cancer-derived tumour spheroids retain characteristics of original tumours [4,5] and lung cancer spheroids their chemosensitivity [6] but cardiac pluripotent cells do adapt to growth in 3D culture [7]. Immortal cell lines recover ('re-differentiate') structural and functional features of their parental tissue and regain in vivo drug sensitivity (e.g., breast, MCF-7 [8,9]; pancreatic β-cell β-TC6 [10] and RIN 5F [11]; glial-like GL15, neuronal-like SH-SY5Y [12]; ovarian, OV-MZ-6 and SKOV-3 [13]; trophoblast BeWo, Jeg-3 and JAr [14]; and liver, HepG2 [15], HuH-6 [16], HepG2/C3A [17–19]. Stem cells, whether embryonic or induced pluripotent, differentiate into 3D organoids when provided with appropriate molecular guidelines (e.g., ECM and growth factors) [20,21]. These recapitulate differentiation into a wide variety of mimetic tissues. (e.g., optic cup [22], pancreas [23] & gastric [24]). Growth in 3D microenvironments boosts the induction of pluripotency [25]. Transplantation of 3D structures into animals induces them to differentiate further, in some cases, to tissues that are almost indistinguishable from the native organ [26].

While the term 'spheroid' is usually used to indicate a mimetic tissue that is constructed from immortal cells and the term 'organoid' used to indicate a mimetic tissue derived from primary cells (including stem cells of any derivation) the basic principles driving recovery will be the same and so for the purposes of this article, the word 'spheroid' is used for both.

Clinostat rotating vessels (Figure 1, also known as rotating wall vessels (RWV), rotating cell culture systems (RCCS) or high aspect rotating wall vessels (HARV)), are commonly used to generate a 'microgravity' environment that is conducive to the production of highly reproducible long lived 3D cultures which allow for the investigation and manipulation of mimetic tissues. Strictly speaking, they produce omnidirectional gravity—i.e., the tissue is influenced by gravity from all sides, effectively neutralizing directional gravity effects. In practice, the gravity is slightly larger than 1G: a clinostat running at 20 rpm will generate a G-force of 1.0089 at 2 cm from the axis of rotation.

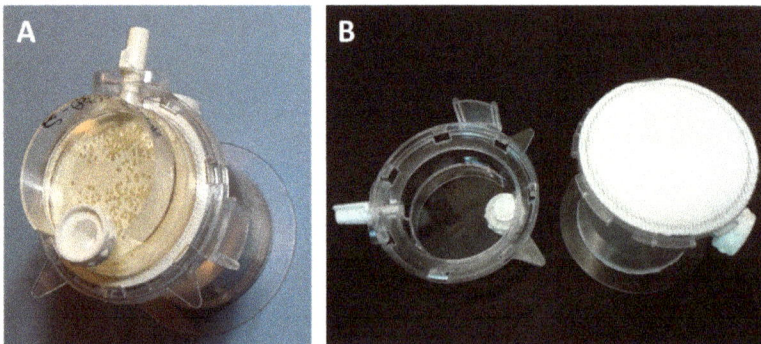

Figure 1. (**A**) an assembled bioreactor containing >300 21 day old spheroids. (**B**) The open bioreactor with (left) a 10 mL petri-dish like culture chamber and (right) the gas exchange membrane. Behind the membrane is a water reservoir and humidification labyrinth. White stoppers allow access for media change or filling the reservoir. This type of bioreactor has a gas membrane exchange area of 13.2 cm^2 and a fixed volume (nominally 10 mL) and is available from CelVivo (Denmark).

The clinostat bioreactor has a number of advantages. Most cells in tissues, with the obvious exception of endothelial cells, experience little or no shear forces. Critical/lethal shear stresses for different mammalian cell types are in the range of 0.3–1.7 Pascal (1 Pa = 10 dyne/cm^2) (Croughan and Wang, 1991) [27].

Mimetic tissue culture in a clinostat bioreactor provides very low shear forces (at 20 rpm, ca. 0.01 Pa [28] on the suspended spheroids, similar to rocking platform 'wave' bioreactors (set at an oscillation of 7° at 20 rpm) [29,30] and 0.02–0.064 Pa for micro-fluidic devices (flow rate 650 μL/min) [31,32].

Higher shear forces (and cellular effects) are seen for stirred suspension bioreactors (100–200 rpm, 0.3–0.66 Pa) [33,34] and for orbital shakers (20–60 rpm, 0.6–1.6 Pa) [35,36].

In rocking platform bioreactors and micro-fluidic devices, the mimetic tissue is usually in contact with plastic surfaces and the shear forces vary either with time, location, or both. This will induce differential growth rates, sizes and biochemical properties of the spheroids [37]. These differences will affect, for example, drug response [38]. The clinostat bioreactor exposes all spheroids to an equal and very low shear force and has been show to result in uniform spheroids that even after 21 days in culture have a standard deviation in their size of only ±21%. Clinostat spheroids are therefore the best suited for studies of metabolism or pharmacology, especially for kinetic measurements.

This recovery of the physiological phenotype in 3D culture suggests that either there is a common driving mechanism or that it occurs spontaneously. Strangely enough, clues as to why this happens can be found in ideas that have been around for almost 100 years.

1.1. The Relationship between Oxidative Phosphorylation and Aerobic Glycolysis

Rapid cancer cell proliferation favours reprogramming from oxidative phosphorylation to aerobic glycolysis. This concept was described initially by Warburg in the 1920's [39,40]. He observed that proliferating ascites tumour cells convert most of their glucose to lactate and referred to this process as aerobic glycolysis because it occurred regardless of whether oxygen was present or absent [41].

In dynamic nuclear polarization (DNP) spectroscopic techniques, hyperpolarized [^{13}C]-labelled pyruvate or other glycolytic intermediates have shown that tumours in situ produce lactate at levels which correlate with their degree of tumour progression or response to treatment [42,43]. Similar studies using hyperpolarized [^{13}C]-labelled glucose revealed increased lactate production in mouse lymphoma and lung tumours, but not in healthy tissues [44]. Other whole-body approaches, such as PET and MRI brain-imaging techniques, have measured that 10–15% of the glucose that is used by the healthy brain is metabolised by aerobic glycolysis [45]. Similar results have been obtained in vitro using perfused heart and liver tissues [46] and in cell culture, comparing non-proliferating myocytes with proliferating Rh30 cell line. The proportion of aerobic glycolysis to oxidative phosphorylation varies between different tumour types: rapidly growing tumours tending to utilise a higher degree of aerobic glycolysis, in keeping with their reduced reliance on oxygen and their need to synthesise larger amounts of precursors faster [40,47].

Warburg hypothesized that metabolic reprogramming was specific to cancer cells, and that it arose from mitochondrial defects [39]. While his observations have been corroborated many times, the hypothesis that cancer growth was driven by these defects has been disproven [40,48,49]. Weinberg et al. demonstrated that mitochondrially generated reactive oxygen species are essential for Kras-induced proliferation and tumourigenesis of HCT116 colon cancer cells [50].

The current viewpoint is that, in aerobic glycolysis, glucose is metabolised via the glycolytic pathway to produce lactic acid, nucleotides, amino acids, and other metabolites. Simultaneously, glutamine is converted via glutaminolysis to citrate for cholesterol and lipid production. In contrast, non-proliferating, differentiated cells in healthy tissues efficiently produce ATP through oxidative phosphorylation. In oxidative phosphorylation, glucose is metabolised via the glycolytic pathway and then the pyruvate produced is oxidised to CO_2 and water in the tricarboxylic (TCA) cycle. The electrochemical gradient generated is used to produce ATP [41].

1.2. Are Growth Rates Inversely Related to Functionality?

Although generally accepted, there is a flaw in the argument that rapidly growing cells preferentially utilise aerobic metabolism. Most tumours and healthy tissues proliferate relatively slowly, doubling their numbers every 20–60 days [51]. In contrast, cultured cells double every 1–8 days and therefore would be expected to exhibit strong aerobic metabolism. This is not the case. In a painstaking review, using oxygen consumption and lactate production to define oxidative and glycolytic ATP production, Zu and Guppy showed that there is no evidence indicating that cancer cells or cells in culture are inherently glycolytic [52]. Despite considerable variation, both normal and cancer cells produce about 19% of their ATP using aerobic glycolysis and the rest by oxidative phosphorylation. They concluded that cancers tend to be glycolytic because they are hypoxic [52]. Metabolic reprogramming to aerobic glycolysis has been called a hallmark of cancer [41,53,54], but in practice, it is a consequence of hypoxia.

There is another paradox. By definition, healthy tissues exhibit full physiological functionality. Tumours tend to lose these functionalities in a reciprocal proportion to their proliferation rate [51,55,56]. Immortal cell lines are often considered a 'terminal condition' for tumour cells where they proliferate rapidly but have lost much of their in vivo functionality. However, when immortal cells are given sufficient time to adapt to 3D conditions, their proliferation rate slows to that seen in tumours and healthy tissues in vivo and they regain physiological functionality [18,19].

So, the question becomes how do growth rate, metabolic reprogramming, and physiological functionality relate to each other. Is proliferation linked to metabolic reprogramming or is it independent? Is there a 'spectrum' between, on the one extreme, healthy tissue and at the other, immortal cell lines grown in 2D? Or are the 'axes' of normal to transformed independent of the axis of hyperoxic to hypoxic? Where should 3D cultures 'be placed' in this spectrum? Does metabolic reprogramming to aerobic glycolysis invariably lead to the loss of physiological function, as seen in the transformation of normal tissue to cancer, or are these phenomena independent? Can cells reprogram between oxidative phosphorylation and aerobic glycolysis based on their growth requirements or does it have consequences? We have addressed these questions by reviewing what is known about metabolic reprograming.

2. Materials and Methods

This manuscript is based on the deeper evaluation of raw data published previously as supplementary data [57]. For convenience, we describe here in brief the methods used. A full description can be found in our previous manuscript.

2.1. Cell Culture

HepG2/C3A cells were grown in DMEM containing 1 g glucose/L. In 2D conditions, they were left until they were nearly confluent (day 5) before they were collected for mass spectrometry. HepG2/C3A cell spheroids were prepared using AggreWell™ 400 plates (Stemcell Technologies, Vancouver, Canada) and left to mature in a rotating 'microgravity' micro-bioreactors on a BioArray Matrix drive (CelVivo IVS, Blommenslyst, Denmark) for at least 21 days to reach a dynamic equilibrium [57]. The clinostat 3D culture reduces shear forces on the cells to a minimum while increasing nutrient and gas exchange (see Supplementary video).

2.2. Determination of Protein Content of Spheroids

Spheroids were washed with PBS, collected, photographed to calculate their individual protein content using a look-up table derived, as described previously [17].

2.3. Determination of Glucose and Glycogen Content of Spheroids

Glucose in the media was measured using a Onetouch Vita glucose meter' and test strips, (LifeScan, Inc., Cat Nos. 6407078 and 6407079 respectively, Milpitas, CA, USA). Total glycogen was measured in individual spheroids using a flurometric assay kit (Sigma cat. No. MAK016, Merck KGaA, Darmstadt, Germany). Glycogen is hydrolysed to glucose and measured in a fluorometer. Amounts of glucose in the spheroids prior to glycogen hydrolysis were negligible.

2.4. Mass Spectroscopy

Protein samples were collected from classical 2D cell culture five days after trypsinisation and from 3D spheroid culture 21 days after spheroid culture initiation. The proteins were quantitated, alkylated, digested with trypsin and washed, stable-isotope dimethyl labelled, and electrosprayed into the LTQ Orbitrap Velos (Thermo Scientific, Waltham, MA, USA). Data on 1346 proteins was deemed statistically reliable and was analysed with reference to multiple programs and information sources including MedLine, SwissProt, Kegg, Ingenuity™ and Go protein annotations [57].

3. Results and Discussion

We previously catalogued a plethora of differences between growing HepG2/C3A cells in 2D and 3D conditions. These include changes in the cell architecture (actin, microtubules, intermediate filaments) and metabolism (glycolysis, fatty acid metabolism, cholesterol and urea synthesis, DNA repair, RNA processing, protein folding and degradation, cell cycle arrest, transport around the cell) [57]. In this manuscript we take the analysis of the raw data a step further and use it to describe a coherent model of the driving force behind these differences between 2D and 3D culture. Where the cell type used to construct the spheroids is not defined in this article, the results refer to the raw data generated with HepG2/C3A.

3.1. Adaptation to Growth in 3D Culture

Cells need time to adapt to growing in 3D cultures and implement the changes catalogued above: to slow growth rates, reorganise their cytoskeleton, establish tight junctions, polarity, relocate membrane transporters and secrete tissue specific extracellular matrix components. Establishment coincides with the time needed for spheroids to reach a radius where their core is severely hypoxic. In a passive diffusion culture system (ultra-low attachment dishes or hanging drop) this requires about 8 days. Using NIH 3T3 fibroblasts or HepG2/C3A hepatocytes, the spheroid's radius is about 160 μm [58,59]. In irrigated spheroids, (i.e., where media flows past them in microgravity cultures), this occurs after 18 days (radius is 450 μm) [18].

Evidence for hypoxia-induced metabolic reprogramming to aerobic glycosylation has been provided by reanalysing proteomic studies of cells grown in 2D and clinostat 3D conditions. Protein abundance has been measured in mature, 21 day old spheroids and compared to that seen in 80% confluent, five day old 2D cultures of the same cells (human hepatocellular carcinoma HepG2/C3A) using quantitative proteomics [57] (the raw data is presented in supplementary information) and [60]. For convenience, the reciprocal has been taken of all values below 1 (i.e., where the protein level is lower in 3D than in 2D), and this is indicated with a negative sign. In this way, equally significant changes, for example, a doubling or a halving of the amount of protein would be indicated by '2.00' or '−2.00' respectively.

The central metabolic pathways are illustrated in Figure 2. Enzyme expression levels illustrate a clear increase in glycolytic, glutaminolytic, hexosamine, and pentose phosphate pathways, as well as increased nucleotide, amino acid, and lipid synthesis. In contrast, enzymes of the TCA cycle are essentially unchanged. This is consistent with metabolic reprogramming [58,61]. If enzyme levels can be used to roughly indicate enzymatic activity, then one third of the glucose is metabolised by oxidative phosphorylation, corresponding to that seen in a rapidly growing tumour.

Metabolic reprogramming renders cancer cells susceptible to growth suppression because of their increased dependence on glucose for these anabolic pathways. Tumour cells predominantly express the embryonic M2 splicing isoform of pyruvate kinase (PKM2) [62]. Short hairpin RNA knockdown of PKM2 leads to its replacement by the adult PKM1 form, reverses metabolic reprogramming, increasing oxygen consumption and reducing lactate production and tumourigenicity in nude mouse xenografts [63]. PKM1 and PKM2 switching is regulated by three heterogeneous nuclear ribonucleoproteins hnRNPA1, hnRNPA2 and the polypyrimidine tract binding protein PTBP1 (or hnRNPI). These proteins bind flanking regions around exon 9, and in doing so, promote PKM2 expression [64]. In 3D spheroids, their expression is reduced (hnRNPA1 −2.20; hnRMPA2 −2.29; PTBP1 −1.23) favouring PKM1 and oxidative phosphorylation. However, no PKM1-specific peptides were detected in the original mass spectrometry data making it impossible to differentiate them [57].

Metabolic reprogramming results in the cell becoming increasingly dependent on glutaminolysis for fatty acid synthesis [41,58,61]. The key enzyme, ATP citrate lyase (ACL) shows increased expression in tumours (lung, prostate, bladder, breast, liver, stomach and colon) [65] and in spheroids (Figure 2). Cytoplasmic isocitrate dehydrogenase 1 produces isocitrate from α-ketoglutarate. IDH-1 mutations have been associated with gliomas via nuclear factor-κB activation in an hypoxia-inducible-factor (HIF1-α) dependent manner [66]. Inhibition of the IDH-1 or ACL inhibits A549 metabolism in vitro and, when injected into nude mice, reduces tumour growth [49,58].

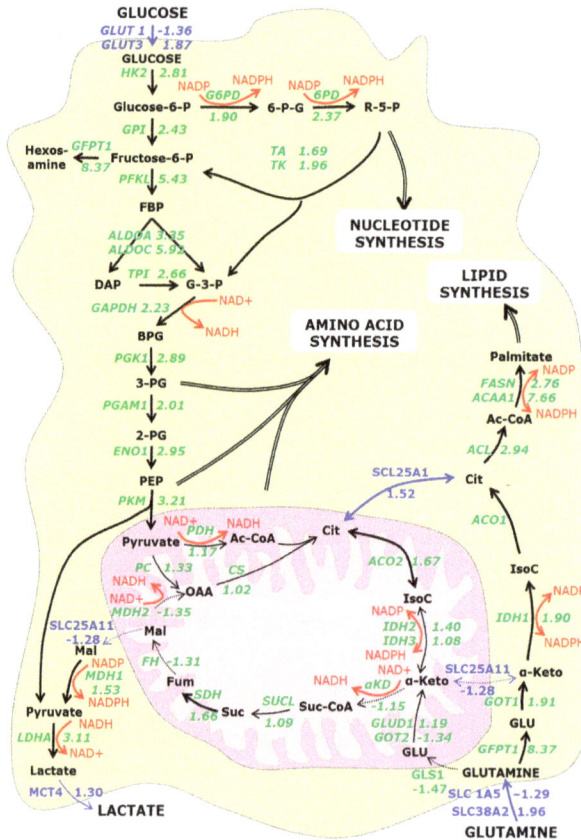

Figure 2. Ratios of the protein abundance of central metabolic enzymes and membrane transporter in 3D compared to two-dimensional (2D) cultures. Metabolites are marked in black, enzymes in green, transporters in blue and selected cofactors in red. Negative ratios indicate that the protein is present in higher amounts in 2D cultures. Arrows connecting metabolites are marked in bold if the enzyme expression is increased by a factor of 1.5 or greater. Arrows connecting metabolites are dotted if the enzyme expression is decreased. (Raw data taken from [57]).

Interestingly, activated effector T cells (TE) utilise anabolic aerobic glycolysis, while memory T cells (TM) use catabolic pathways. Microscopy shows that TE cells have punctate mitochondria, while

TM cells maintain fused networks. The protein Opa1 is required for maintaining fused networks [67]. Opa1 is reduced in 3D spheroids (OPA1 −2.71), also indicating that they utilise aerobic glycolysis [18].

Evidence for metabolic reprogramming of cells in culture has also been found using hyperpolarized [^{13}C] spectroscopy. Two cell lines (Huh-7 hepatocellular carcinoma cells and SF188-derived glioblastoma cells) have been cultivated in 2D and pulse-labelled with hyperpolarized [1-^{13}C] pyruvate to determine the activities of pyruvate dehydrogenase (PDH, as a surrogate indicator of oxphos) and pyruvate carboxylase (PKM, for lactate). While both enzymes were active, supplementation with glucose favoured lactate production. Inhibition of glycolysis using an Akt inhibitor reversed this effect [68]. This illustrates that cells can adapt their metabolic activity to their environment. Rat hepatoma cells (JM1) have been probed using [^{13}C]-labelled glucose while cultivated in either 2D or 3D conditions (encapsulated in alginate beads). These studies showed that in both conditions, 85% of [^{13}C]-glucose was converted either to lactate or alanine by aerobic glycolysis [69].

Jiang et al., compared the metabolic activity of human H460, A549, MCF7, and HT-29 cells grown in 2D and 3D cultures [1-^{13}C] glutamine or [5-^{13}C] glutamine tracers clearly illustrated that citrate and lipids predominantly were synthesised via reductive glutaminolysis. In particular, while neither isocitrate dehydrogenase-1 nor -2 (cytosolic or mitochondrial, respectively) were necessary for monolayer growth, spheroids were dependent on the cytosolic IDH1 for glutaminolysis. Many cell lines (lung, mammary, colon, embryonic fibroblasts, squamous cell carcinoma, melanoma, glioblastoma, and leukaemia) use glutamine as their primary source of acetyl-CoA for lipogenesis [50,70]. Glycolytic ATP is only necessary in hypoxic conditions [37] and glutamine consumption is increased by reducing the available oxygen to 1% [61]. In these conditions, the major role of glucose metabolism is to drive the pentose phosphate pathway to generate NADPH. Glutaminolysis drives acyl-CoA production and lipidogenesis [50].

3.2. Is Metabolic Reprogramming Driven by Oxygen or Glucose Insufficiency?

Given that spheroids constructed from many types of cell exhibit metabolic reprogramming, the question arises as to what causes the switch?

3.2.1. Diffusion Gradients and the Importance of Irrigation

Mammalian cells need oxygen and nutrients. In tissues, they are normally located within 100 to 200 μm of capillaries [71]. This corresponds to roughly 10–40 cell layers thick. An experimentally derived diffusion limit (i.e., where the PO_2 level falls to 0) of 232 ± 22 μm agrees well with this [72]. Many cells in tissues experience low oxygen tension (e.g., 1% O_2) [73] and it is alternative stressors, such as serum deprivation or acidosis, which induce cell death [74]. Only severe hypoxia (<0.01%) O_2 is capable of inducing apoptosis [75].

The most significant differences between 2D and 3D culture are diffusion gradients. Several types exist including gasses, nutrients, metabolites, signalling molecules, secondary messengers and growth factors. Here we will only consider oxygen, CO_2 and glucose, since they have been suggested to drive metabolic reprogramming and need to be taken into consideration when designing micro-bioreactors (Figure 3).

The existence and depth of the hypoxic zone depends on several factors, including radius, cell type, and media flow rate. Measured diffusion gradients for oxygen follow smooth sigmoidal curves (with no difference in shape outside or inside EMT6 spheroids), suggesting that the presence of cells has little influence on its diffusability. Atmospheric oxygen (21%) provides a partial pressure (PO_2) of about 145 mm Hg (ca. 190 μM) in media. In static cultures where there is no flow of plasma or media, the PO_2 falls rapidly towards the spheroid's centre (Figure 3A). Small radius spheroids (25–50 μm) have about 3.3% PO_2 in their core, close to physiological levels in the brain. 100 μm radius spheroids have about 1.6% PO_2 and larger spheroids less [76]. In irrigated spheroids, where media flows past the spheroid, the PO_2 falls to about 13% measured in the media at the surface of the spheroid [77]. Thus, even when irrigated, there is a 'diffusion-depleted zone' in the media surrounding each spheroid (grey zones in Figure 3). The PO_2 reaches a minimal plateau of 3.3% at about 150 µm into a 352 µm radius irrigated EMT6 spheroid, and 1.6% at about 225 µm into a 480 µm radius spheroid. In the latter case, stopping the media flow causes the core PO_2 to quickly fall to 0. Doubling the flow rate had a marginal effect and was confined to the spheroid surface [78]. Core PO_2 reached 0% in irrigated spheroids with radii greater than 600 µm.

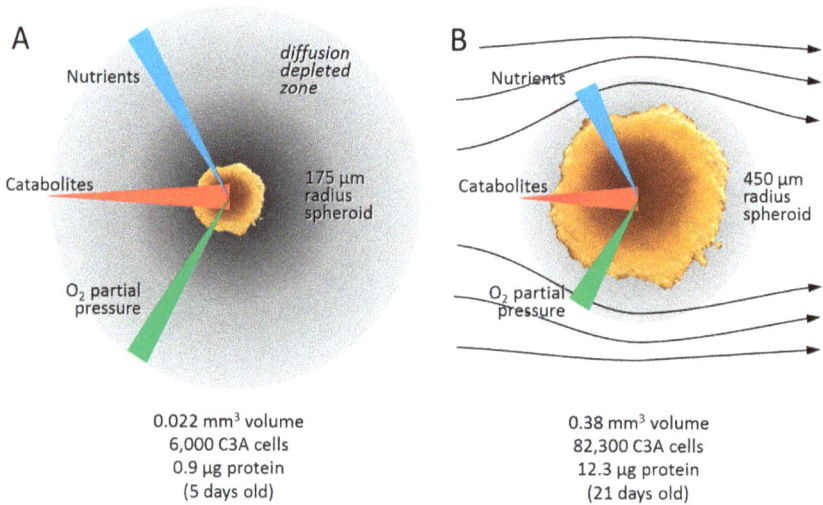

A

Nutrients

Catabolites

diffusion depleted zone

175 µm radius spheroid

O_2 partial pressure

0.022 mm³ volume
6,000 C3A cells
0.9 µg protein
(5 days old)

B

Nutrients

Catabolites

450 µm radius spheroid

O_2 partial pressure

0.38 mm³ volume
82,300 C3A cells
12.3 µg protein
(21 days old)

Figure 3. The diffusion depleted zone and its consequences for the oxygen, nutrient and catabolite gradients in (**A**) passive and (**B**) irrigated 3D culture. The size shown indicates the approximate maximum size above which anoxia develops in the spheroid core. Volumes, number of cells and amounts of protein are indicated for each. The apparent increase in cell volume is attributed in part to increased ECM and the development of sinusoidal and bile cannalicular spaces between the cells.

Haemoglobin is normally found in hepatocytes and several cell lines. Hepatocarcinoma spheroids increase their haemoglobin content by a factor of thirty to actively alleviate low PO_2 oxygen concentrations [57].

Experiments using pH as a spatial readout have demonstrated that the diffusivity of CO_2 through spheroids (colorectal HCT116 and HT29, breast MDA-MB-468, pancreatic MiaPaca2, cervical squamous cell carcinomas HeLa and SiHa and ovarian clear cell adenocarcinoma OVTOKO) is exactly the same as its diffusivity through water (2.5×103 µm²/s) [79]. Usually, the spheroids' cores are slightly more acidic than the surrounding media, possibly due to increased CO_2 or lactate amounts [80,81].

No data is available describing a glucose gradient in or around spheroids. When considering that the glucose molecule is larger than oxygen or CO_2, the gradient would intrinsically be expected to be steeper, but glucose transporters may alleviate this.

3.2.2. Hypoxia Affects Glycolysis and Oxidative Phosphorylation

Hypoxia has numerous effects on mammalian cells. One is the activation of the constitutively expressed hypoxia-inducible factor (HIF-1α) (Figure 4).

Under normoxia, defined as the PO_2 levels normally seen in healthy tissues (usually 1–5% [73]), HIF-1α subunits are hydroxylated by prolyl hydroxylases (PHD1-3 including the TCA cycle enzyme α-KD). The modified HIF-1α is recognized and targeted for proteasomal degradation by the VHL-E3-ubiquitin ligase complex.

When oxygen concentrations decrease, the oxygen-dependent PHDs are inactivated, allowing for the HIF-1α protein to accumulate. This promotes HIF-1α translocation to the nucleus where it interacts with HIF-1β/ARNT and p300. This complex binds hypoxia-response elements (HREs) in promoter regions of numerous target genes, including glucose transporters and glycolytic enzymes [82].

The translationally controlled tumour protein (TCTP 4.49) binds competitively to VHL, reducing PHD binding and accelerating HIF-1α accumulation, nuclear translocation, and transcription reprogramming [83].

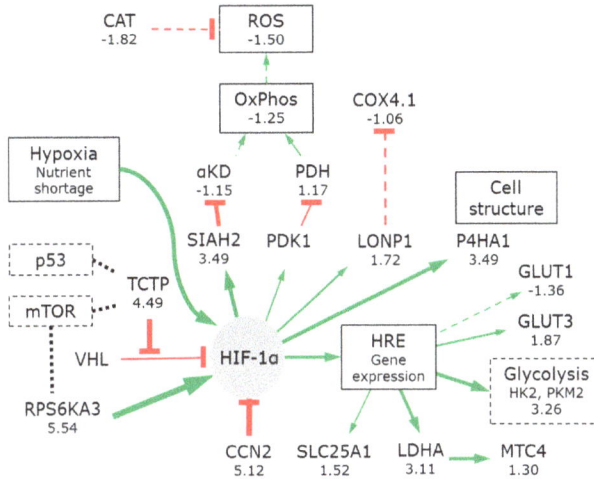

Figure 4. Effects of hypoxia on HIF-1α. The thickness of the line indicates the ratio of protein amount in thre—dimensional (3D) cultures compared to 2D cultures. Dotted lines indicate reduced expression. Green lines ending in arrowheads indicate activators, while red lines ending in a bar indicate inhibitory activity. Dotted-boxes indicate links to other pathway figures.

HIF-1α induces pyruvate dehydrogenase kinase 1 (PDK1) expression. PDK1 inhibits the mitochondrial pyruvate dehydrogenase (PDH) [84]. This reduces pyruvate flux into the TCA cycle and lowers the mitochondrial oxygen requirements. This switch increases lactate production and secretion, as observed by Warburg. Differentially transformed rat embryo fibroblasts showed increasing levels of lactate content and unchanged or decreasing lactate secretion in irrigated spheroids with increasing radii of up to about 450 μM. Above this radius, lactate content and secretion stabilised, illustrating that hypoxia-induced glycolysis need not lead to lactate secretion [58,81] suggesting that most of the glycolytic metabolites are utilised in anabolic processes.

Liver cells can convert lactate back to pyruvate. Despite this, the lactate transporter (MCT4 or SLC16A3) is increased (1.30) suggesting that the cells might 'pump' the lactate towards the spheroid surface.

Spheroids show the increase in glucose transporters, glycolytic enzymes and lactate dehydrogenase by on average about a factor of 3.26. This metabolic reprogramming is partial: spheroids do not show a significant decrease in the pyruvate dehydrogenase or of any enzymes of the TCA cycle (Figure 2).

HIF-1α also induces E3-ubiquitin ligase SIAH2 synthesis. This mediates the proteasomal degradation of the OGDH subunit of α-KD and forms part of the feedback control of HIF-1α. A modest reduction of the α-KD 3 enzyme complex is observed in spheroids (DLD −1.26, DLST −1.11, OGDH −1.09). This will slow the TCA cycle and allow for more citrate to be transported into the cytoplasm by an upregulated citrate transporter protein (SLC25A1, 1.52), supporting the metabolomics [81] and isotope analyses [58,61].

Interestingly HIF-1α also promotes extracellular matrix remodelling via collagen hydroxylases (P4HA1 3.49), a facility useful for cancer cell morphology, adhesion, and motility [85].

Part of the indirect negative feedback regulatory circuit for HIF-1α is the connective tissue growth factor (CCN family member 2) or insulin-like growth factor-binding protein 8, (IBP-8). It is strongly upregulated in spheroids (5.12) illustrating strong positive and negative regulatory mechanisms are active.

HIF-1α can also induce the mitochondrial protease LONP1, which degrades the less efficient cytochrome C oxidase 4 subunit 1 (COX4-1) from the complex IV of the electron transport chain and allows it to be replaced by the more efficient COX4-2 [82]. Although LONP1 was increased (1.72), there was no change in the level of COX4-1 (−1.06). LONP1 is an essential central regulator of mitochondrial activity and is overexpression in oncogenesis [86]. Despite that spheroids contain higher levels of ATP, subunit IV and the ATP synthase (subunit V) are reduced by −1.12 and −1.37 respectively [18,19]. Reduced mitochondrial respiration will result in fewer reactive oxygen species correlating with reduced levels of catalase (CAT −1.82) [87] resulting in diminishes hydrogen peroxide damage and 50% less oxidised proteins.

3.3. Glucose Starvation Has Little Effect on Metabolic Reprogramming

The feature that Warburg noticed—that cancer cells rapidly use glucose and convert it to lactate would suggest that glucose availability might also play a central role.

Liver cell spheroids are known to rapidly import glucose and convert it to glycogen. When cultured in bioreactors with physiological amounts of glucose (5.5 mM), the media glucose is typically exhausted 8 h after media exchange (Figure 5). Thereafter, the spheroids experience 'glucose starvation' and catabolise the glycogen they have synthesised.

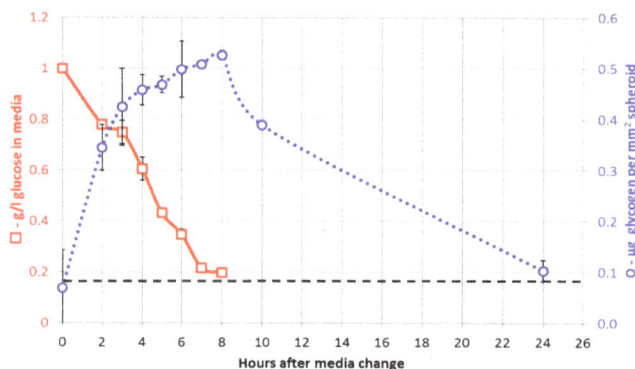

Figure 5. Relationship of glucose consumption from the media with the level of glycogen present in the spheroids. The amount of glucose in the spheroids prior to glycogen hydrolysis was negligible.

Glucose starvation could unleash a number of changes in the cell, as initiated by the Glucose Regulated Proteins (GRP). These are typically found in the ER, often overexpressed in cancers and associated with aggressive growth and invasion [88]. Their first effect would be to increase ER stress and initiate the unfolded (or misfolded) protein response (UPR) [89]. In the UPR, GRP78 dissociates from three protein-folding quality sensors (IRE1, PERK, and ATF6) embedded in the ER membrane. These sensors activate the UPR signal transduction program, a negative feedback loop that alters gene expression to slow protein synthesis and the cell cycle (eventually leading to arrest in G1 [90,91]).

Surprisingly, the amount of GRP78 is unchanged between 2D and 3D (1.05) and there are only weak changes in GRP58 (1.34) and GRP60 (1.37), suggesting that there is no UPR or ER stress.

The mitochondrial GRP75 can inactivate p53 and induce apoptosis [88], but it is only slightly elevated (1.36), suggesting that the mitochondria also suffer very little stress.

GRP94 and GRP170 show the strongest responses (2.09 and 2.15, respectively). GRP 94 plays critical roles in folding and exporting proteins in the secretory pathway (e.g., insulin-like growth factors IGF-1 and 2), which could activate the PI3K-Akt pathway. GRP170 is a glycosylated protein also known as the hypoxia up-regulated protein 1, HYOU1. It plays a role in suppressing apoptosis and is up-regulated in invasive tumours [88]. Considering the relatively little stress caused by prolonged glucose starvation, the lack of glucose appears to play a minor role in the metabolic reprogramming. What effects there are, appear to stabilise cellular metabolism and are anti-apoptotic.

3.4. Metabolic Reprogramming 'Links' Glutamine Metabolism to the Hexosamine Pathway

Metabolic reprogramming results in an increased reliance on glutamine. Intracellular levels are regulated by plasma membrane transporters SLC1A5 and SLC38A2 [92]. ER stress would induce their degradation and ultimately to autophagy and cell death [92,93]. In spheroids, SLC38A2 is increased (1.96) while SLC1A5 is decreased (−1.29), suggesting that they play subtly different roles. In agreement with this, net glutamine uptake in HeLa cells was not dependent on SLC1A5 but required SLC38A1 or 2 [94].

3.4.1. Conversion of Glutamine to Glutamate

The cells' glutamate demand is probably supplied by the highly upregulated GFPT1 which is the first, and rate-limiting step, of the hexosamine pathway (8.37). This enzyme catalyses the conversion of fructose 6-phosphate and glutamine to glucosamine 6-phosphate and glutamate. Activation of glutaminolysis was necessary for adaptive cell survival in the mouse model of pancreatic ductal adenocarcinoma [95]. Hypoxia is considered to drive this adaptive process, which, amongst other things, leads to increased amounts of O-linked N-acetylglucosaminylated proteins. In agreement with this, several polysaccharide, proteoglycan and glycosylation synthetic pathway enzymes are strongly upregulated in 3D spheroids (UDP-glucose pyrophosphorylase UGP2 6.59; UDP-glucose 6-dehydrogenase UGD 7.46; UDP-glucose 4-epimerase GALE 12.71; and sialic acid synthase NANS 5.37).

3.4.2. α-Ketoglutarate

Glutamate can be converted to α-ketoglutarate by the mitochondrial GLS1 (−1.47) and GLUD (1.19). It can also be converted by cytoplasmic or mitochondrial alanine or aspartate aminotransferases [50] and it is the cytoplasmic enzyme that is upregulated (GOT1 1.91, GOT2 −1.34).

There are three possible routes by which α-ketoglutarate can be converted to citrate (Figure 2). Firstly, it can be converted around the TCA cycle. Secondly it could be converted via isocitrate to citrate (by IDH2 1.40 and ACO2 1.67), reversing the normal TCA cycle flux by reductive glutamate metabolism [61]. IDH3 is not increased (1.08) because it can only catalyse the 'forward' reaction. Finally cytoplasmic α-ketoglutarate, produced via the upregulated GFPT1 and GOT1, can be converted, by IDH1 (1.90) and ACO1. While all three processes probably occur, both isotope tracing and enzyme abundance suggests that the latter route is the most active [57,61].

3.4.3. NADH

Conversion of α-ketoglutarate to isocitrate requires the cofactor NADPH. The reduction in the mitochondrial MDH2 (−1.35) and the essentially unchanged abundance of its NAD(P) transhydrogenase (NNT 1.13) suggest that the mitochondrial source is of low significance. MDH1 is increased (1.53) but lacks a malate source (the SLC25A11 transporter is reduced −1.28) and the conversion of cytoplasmic pyruvate to lactate would actually consume the NADH that is produced. The richest source of NADPH is the pentose phosphate pathway where G6PD and 6PD are both upregulated (1.90 and 2.37 fold respectively), in agreement with isotope tracing data [57,61].

3.4.4. Citrate

Citrate is used for fatty acid synthesis. ATP-citrate synthase (ACL 2.94) uses citrate to generate cytosolic acetyl-CoA. Acetyl-CoA is used for: histone acetylation by acetyl-CoA acyltransferase (ACAA1 7.66); palmitate synthesis by fatty acid synthase (FASN 2.76); cholesterol, steroid hormones, haem and a plethora of other biomolecules. Glutamine is as important as glucose in metabolic reprogramming and blocking glutamate-dependent cellular pathways (at either IDH1 or ACL) limits tumorigenic growth [49,58].

3.5. Metabolic Reprogramming Is Associated with Chromatin Remodelling

Conversion between transcriptionally active euchromatin and inactive heterochromatin is brought about by processes, including acetylation, methylation, and clipping of histones. Hypoxia can change these epigenetic markings. HIF-1α stabilisation leads to increases in histone lysine demethylases (KDM3A, KDM4B, KDM4C, and KDM6B) [96]. While C3A cells that are grown in 2D culture essentially show little epigenetic marking, spheroids recover extensive histone methylation, acetylation, and clipping on both H2B and H3 [97]. Hypoxia also upregulates the arginine *N*-methyltransferase PRMT1 (2.66), increasing methylation of arginine 3 of H4 [98]. PRMT1 can asymmetrically methylate the ReIA subunit [99] inhibiting its binding to DNA and repressing NF-κB target genes. The 'Chromatin target of PRMT1' protein, (CHTOP) [100] which promotes cell cycle progression, is strongly reduced in spheroids (−8.09), resulting in few cells in the G2/M phase. Histone deacetylases do not appear to be affected (HDAC1 1.00).

3.6. The Switch to Anabolic Metabolism

Spheroids composed of either OVTOKO or SIHA cell lines have been shown to contain higher levels of serine, glutamine and other amino acids as well as citrate [81]. The amounts of all anabolic rate-limiting enzymes are increased while catabolic enzymes are unchanged (Table 1) in concordance with metabolic reprogramming to aerobic glycolysis.

Table 1. Rate limiting enzymes for central metabolic pathways and the ration of their expression in 3D spheroids compared to 2D exponential growth. n.d. not detected.

Pathway	Gene	Fold Change
Glucose phosphorylation	HK2	2.81
Glycogenolysis	PYGB	4.06
Glycolysis	PFKL	5.43
Glycolysis	PKM	3.21
Pentose Phosphate	G6PD	1.90
Hexose	GFPT1	8.37
TCA Cycle	IDH2 & 3	1.40 & 1.08
Pyrimidine synthesis	CAD	3.49
Purine synthesis	PRPS1	3.74
Fatty acid synthesis	FASN	2.76
Fatty acid synthesis	ACAA1	7.66
Fatty acid oxidation	CRAT	1.17
Alanine synthesis	ALT	n.d.
Asparagine synthesis	ASNS	5.58
Aspartate synthesis	GOT1	1.91
Cysteine synthesis	MAT1	5.62
Glutamine-glutamate conversion	GLUD	1.19
	GFPT1	8.37
Glycine synthesis	SHMT	1.47
Methionine synthesis	MTR	n.d.
Proline synthesis	PYCR1 & 2	1.06 & 1.03
Serine synthesis	PHGDH	7.67
Tyrosine synthesis	PAH	3.80
Urea synthesis	CPS	3.49
Folate synthesis	MDHFD1	2.49

The three rate-limiting glycolytic pathway steps (HK2, PFKL, and PKM) are three of the four most increased enzymes of the pathway (the 4th being aldolase). Interestingly, PFKL is repressed by high ATP/AMP ratios [101]. Since spheroids have high ATP amounts [19], high PFK levels suggest that AMP levels are also high.

The glutamine-dependent cytosolic carbamoyl-phosphate synthetase 2, is upregulated in spheroids (CAD 3.49). CAD is the rate-limiting enzyme carrying out the first three steps in pyrimidine synthesis. CAD is essential for uridine diphosphate (UDP) synthesis, which in turn, is essential for glycogenesis. This correlates with the appearance of glycogen granules in hepatocyte spheroids and with protein glycosylation and the hexosamine pathway [102].

3.7. Signal Pathways Involved in Orchestrating Metabolic Reprogramming

All of the adaptations seen in glycolysis and glutaminolysis, pentose phosphate pathway, TCA cycle, and fatty acid synthesis indicate that spheroids, grown in a wide variety of 3D culture systems, are utilising a significant degree of metabolic reprogramming to aerobic glycolysis.

The typical features of 3D culture—diffusion gradients resulting in hypoxia (and to a less extent glucose starvation) clearly drive metabolic reprogramming. Warburg saw this phenomenon as a hallmark of cancer. In order to investigate how metabolic reprogramming is orchestrated, we reviewed the status of pathways that are often associated with tumour development: PI3K/Akt/mTOR, Myc, p53, nuclear factor kappa-B (NF-κB), and Wnt [54].

3.7.1. PIK3/AKT/mTOR

The PI3K/AKT/mTOR pathway (Figure 6) plays a key integrating role, sensing concentrations of nutrients (including glucose, oxygen, amino acids and ATP levels) and regulating the anabolic processes of the cell for growth and maintenance [103].

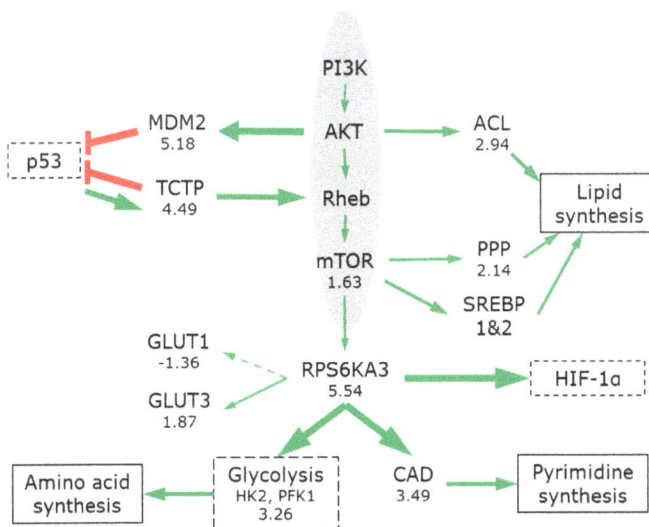

Figure 6. mTOR signalling in 3D spheroids. See legend to Figure 4 for nomenclature.

While only two key proteins from this pathway were detected (mTOR, 1.63; and ribosomal protein S6 kinase RPS6KA3, 5.54), strong downstream effects are clearly visible showing that pathway is activated in 3D (Table 1). mTOR signalling increases translation of hypoxia-inducible factor 1α (HIF-1α), glucose transporters and glycolytic enzymes, and promotes metabolic reprogramming [101,104] (Figure 6).

mTOR promotes pentose phosphate pathway (PPP) enzyme expression (on average by 2.14) and channels metabolic flux into its oxidative, NADPH-producing branch [91]. mTOR strongly stimulates pyrimidine synthesis via the RPS6KA-mediated phosphorylation of CAD (3.49), thereby increasing the pool of nucleotides available [105]. AKT can phosphorylate ACL, enhancing its lipogenic activities and mTOR signalling promotes NADPH-requiring lipid synthesis by activating sterol regulatory element-binding proteins (SREBP1 and 2) [106].

3.7.2. Myc

Myc has the potential to play a key role in metabolic reprogramming. Myc is central to growth regulation and is one of the most frequently deregulated oncogene transcription factors seen in a wide variety of cancers [107,108]. Myc directly transactivates gene expression of GLUT1, phosphofructokinase (PFK), enolase (ENO) and LDHA and indirectly increases phosphoglucose isomerase (GPI), glyceraldehyde-3-phosphate dehydrogenase (GAPDH) and phosphoglycerate kinase (PGK1) [109] (Figure 7). This is consistent (with the exception of GLUT1) with their increased levels in spheroids. However, as described above, PIK3/AKT/mTOR can also induce these proteins (via HIF-1α) and so this effect need not be attributed to Myc. HIF-1α can inactivate Myc [110], and in doing so, induce cell cycle arrest [111].

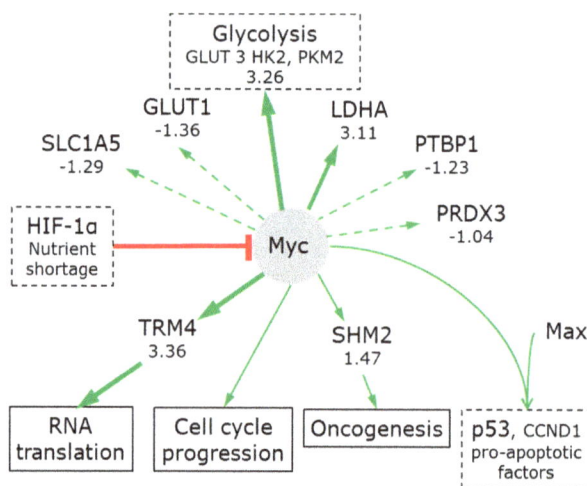

Figure 7. Myc signalling in 3D spheroids. See legend to Figure 4 for nomenclature.

Low expression levels of several proteins normally induced by Myc suggest that Myc is not particularly active in 3D spheroids. Examples include: PTBP1 [112] (−1.23); GLUT1 (−1.36); SLC1A5 [107] (−1.29); and, PRDX3's [113] (−1.04). Myc regulates serine hydroxymethyl transferases and pathway hyperactivation is a driver of oncogenesis [107]. However, the moderate increase of SHMT2 (1.47) cannot qualify as hyperactivation. One exception may be tRNA (cytosine34-C5)-methyltransferase (TRM4 which methylates the first position of the cytosine anticodon). Myc enhances TRM4 expression (3.36). The formation of a covalent complex between dual-cysteine RNA:m5C methyltransferases and methylated RNA has been proposed to provide a unique mechanism by which metabolic factors can influence RNA translation, in particular the processing and utilisation of m5C-containing RNAs [114].

Nutrient shortage and/or hypoxia can inhibit Myc translation; reduce its stability and its ability to dimerise with another transcription factor MAX. Inhibition of Myc/MAX dimerization prevents specific gene expression, most significantly of p53, cyclin D1 and pro-apoptotic factors [115]. Therefore, while Myc regulates many proteins in cancer [107], it appears that the slow proliferation of cells in spheroids is a result of low myc activity.

3.7.3. p53

The tumour suppressor p53 can transactivate a broad array of target genes that are involved in redox maintenance, DNA repair, cell cycle checkpoints, and can thus affect cellular senescence, proliferation, and apoptosis. Mutations of p53 are found in over 50% of human tumours and disturb the IGF1-AKT branch of the mTOR pathway [116].

p53 activity is tightly linked to the oncogene protein DJ-1 (1.83). In a self-regulating loop, DJ-1 is necessary for hypoxic stress-induced p53 activation, while p53 prevents the accumulation of the DJ-1 protein (Figure 8). DJ-1 can bind the ubiquitin-independent 20S proteasomal core and its quantitative increase mirrors the increase in the core (1.92) and in NADPH:quinone oxidoreductase 1 (NQO1, by 2.42) (which protects p53 from proteasomal degradation).

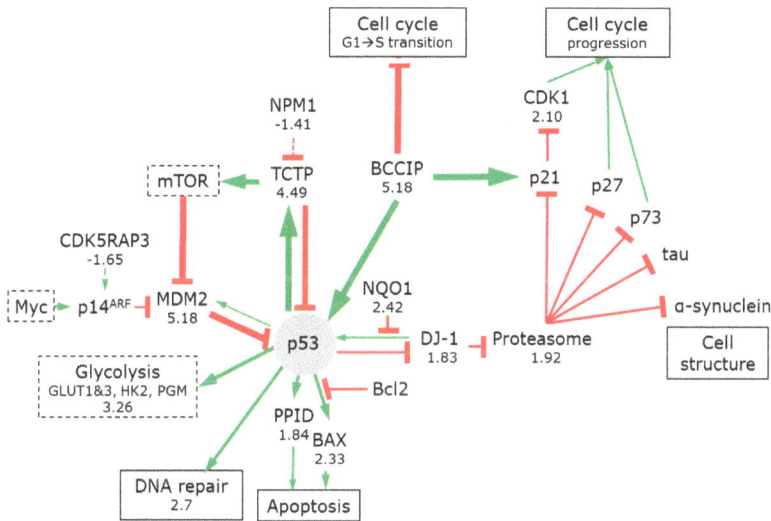

Figure 8. p53 signalling in 3D spheroids. See legend to Figure 4 for nomenclature.

Many key regulatory proteins, including tumour suppressors p53 and p73, tau, α-synuclein and the cell cycle regulators p21 and p27 are degraded by the proteasome core. DJ-1 binding inhibits the activity of the core and by slowing their degradation, leads to an increase in their abundance and activity [117]. DJ-1 can also activate the AKT/mTOR pathway [116].

In addition, the transcriptional suppressor CDK5RAP3 is reduced (−1.65). This will allow for the synthesis of p14ARF and its binding to MDM2. This releases p53 from inhibition. This results in the stabilization, accumulation, and activation of p53 [118].

Activation of p53 is consistent with the increase in DNA repair enzyme expression [57] (on average by 2.7). Interestingly, both the positive (BCCIP) and negative (TCTP) p53 regulators are strongly increased (5.18 and 4.49, respectively), illustrating that p53 is subjected to a tight feedback regulation. BCCIPβ plays a role in cell growth regulation [119]. Overexpression of the BCCIPβ splices variant delays the G1-to-S cell cycle transition and elevates p21 expression. Elevated p21 expression would inhibit cyclin dependent kinase 1 (CDK1 2.10) induction of cell cycle progression.

The evolutionarily conserved TCTP is emerging as a pleiotropic key to phenotypic reprogramming through its ability to regulate the mTOR pathway [120], as well as being an upstream activator of OCT4 and NANOG transcription factors (which play essential roles in nuclear reprogramming). p53 induces TCTP, reducing oxidative stress and minimizing apoptosis [121]. Forming another negative feedback loop, TCTP can inhibit both transcription and function of p53 [122]. The activation of TCTP (4.49) suggests reduced proliferative drive [123]. This is confirmed by the reduction in nucleoplasmin (NPM1 −1.41), which would otherwise complex with TCTP during mitosis to promote cell proliferation.

p53 directly regulates cellular redox homeostasis by modifying expression of pro- and anti-oxidant enzymes peroxiredoxins and thioredoxins. Peroxiredoxins (that act as both sensors and barriers to MAPK activation) are upregulated (PRDX1-6 of 1.74, 2.27, −1.04, 1.86, 1.42, and 2.75, respectively), as is thioredoxin (TXN 1.84). Their upregulation contributes to Myc regulation [124]. The modest changes in mitochondrial peroxiredoxins (PRDX 3 and 5: −1.04 and 1.42) suggest that mitochondrial ROS are insignificant 'stress factors' in keeping with the relatively reduced mitochondrial activity.

p53 is activated, but is exposed to tight feedback control. Together with the low activity of Myc, p53 and associated pathways arrest the cells predominantly in G1 or Go [19].

3.7.4. Wnt GSK-3β/β-Catenin

In the canonical Wnt pathway, the Wnt ligand can bind to a Frizzled family receptor, causing a deactivation of the β-catenin destruction complex. This leads to the dephosphorylation of β-catenin, its accumulation and migration to the nucleus where it acts as a coactivator of TCF/LEF transcription factors. Activation of the Wnt/β-catenin pathway activates cell proliferation and the homeostatic renewal of the liver from pericentral hepatocytes [125]. In 3D spheroids, this pathway is inactive: the amount of β-catenin is reduced (CTNNB1 −1.57), and the protein phosphorylase 2A, although present (PPP2RA1 (the constant regulatory subunit core of the PP2A) 1.02), is strongly inhibited by I1PP2A and I2PP2A (5.77 and 2.18). mTOR also negatively regulates PP2A, allowing for the integration of these two pathways. Reverse regulation occurs in amino-acid depleted conditions: PP2A can inhibit mTOR via dephosphorylation of p170 [126].

The histidine triad nucleotide-binding protein 1 is significantly increased (HINT1, 5.99). It keeps the Wnt GSK-3β/β-catenin pathway inactive, limits cell growth [127], and promotes apoptosis via p53 and Bax [128]. The mitochondrial HINT2 may also promote apoptosis (1.32).

In contrast, the non-canonical Wnt pathway appears to be active in 3D spheroids. Binding of Wnt to Frizzled recruits Dsh, which then binds directly to RAC1 (1.40) and indirectly to profilin (2.75) amongst others. Both of these and numerous other upregulated actin-structure modifying proteins lead to the dramatic restructuring seen in spheroids [57].

3.7.5. NF-κB

NF-κB is a rapid-acting primary transcription factor well suited to respond to harmful stimuli like cell stress, cytokines and free radicals. Many different types of human tumours have constitutively active NF-κB [129].

In its inactive state, the NF-κB heterodimer (composed of p50 and ReIA) is complexed with its inhibitor IκBα. In spheroids, HINT1 (5.99) promotes IκBα stability maintaining NF-κB inactive [127]. Hypoxia upregulates PRMT1 (2.66) [57], which asymmetrically methylates ReIA inhibiting ReIA's binding to DNA, and further repressing NF-κB [99]. The type III transforming growth factor β receptor (TGFβR3) regulates both the epithelial-mesenchymal transition and cell invasion during development via NF-κB activation [130]. Deactivation of TGFβR3 (−4.58) is consistent with NF-κB inactivity. The extracellular matrix TGFβ-induced protein (TGFBI), involved in tissue remodelling and found in liver metastases stroma, is very highly upregulated (TGFBI 8.83) [131]. TGFBI reduces NF-κB activation [132]. Deactivation of NF-κB in 3D sensitises the cell to apoptosis or necrosis by allowing for TNF-α to active the JNK pathway and lead to cell death.

3.7.6. Cell Death

Many of the pathways described above influence necrosis and apoptosis. The fundamental difference between them is that bioenergetic failure in necrosis leads to free radical damage, swelling, rupture, and cytolysis, while apoptosis is ATP-requiring and leads to shrinkage, caspase activation, DNA fragmentation, and retention of the plasma membrane [133].

Apoptosis is often 'defeated' as a cell is transformed from healthy to tumourigenic. Many specific mechanisms operating in many organelles can lead to apoptosis [133]. The apoptotic potential is a balance between pro- and anti-apoptotic signals, which are integrated in mitochondria. The decreased amounts of NF-κB and other factors noted above, result in the under expression of anti-apoptotic proteins including Bcl-2; Bcl-XL; NR13; Bcl-2 inhibitor of transcription 1 (PTRH2 −1.37); Bcl-2-associated transcription factor 1 (BCLAF1 −2.61); Bcl-2-associated athanogene 2 (BAG2 −1.18); Bcl-XL-binding protein v68 (PGAM5 −1.78) and the 'defender against apoptotic cell death' (DAD1 −1.37). These anti-apoptotoic proteins would otherwise bind and inactivate pro-apoptotic proteins. The only pro-apoptotic protein detected, BAX (Bcl-2-like protein 4), was increased (2.33). The net result in spheroids is to increase their apoptotic sensitivity, but without activating apoptosis.

Necrosis, as judged by the microscopic appearance of core cells, has often been reported for spheroids. Activated p53 would interact directly with PPID and push the cell towards necrosis [134]. This interaction may be enhanced by increased BAX abundance (2.33), especially when anti-apoptotic Bcl2 proteins are depleted. Thus, both apoptotic and necrotic processes are sensitised.

3D spheroid cultures have illustrated that the serine protease tumour suppressor MASPIN facilitates the mitochondrial permeability transition (MPT). However, since ATP levels are high, neither process opens the MPT pore. Its opening would initiate a collapse of the transmembrane proton gradient and lead to apoptotic or necrotic cell death (depending on the initiating factors). The essential component of the MPT pore, the peptidyl-prolyl isomerase D, located in the mitochondrial matrix is increased (PPID 1.84). The non-essential components, VDAC, (Voltage Dependent Anion Channel, which spans the outer membrane) and ANT (Adenine Nucleotide Translocase which spans the inner membrane) are either unchanged or are decreased (VDAC1 1.02; VDAC2 −1.16; VDAC3 −1.23; ANT1, (ATP/ADP antiporter SLC25A4) −1.01; ANT2, −1.98; ANT3 −1.25). ANT1 can interact with BAX. ANT2 is anti-apoptotic and it's reduction matches other anti-apoptotic BCl-2 proteins. On the balance, necrosis may be favoured over apoptosis due to the reduction in the chromatinolytic activity of AIFM1 (apoptosis-inducing factor mitochondrion-associated 1, −1.90) [135].

4. Conclusions

The most widely used approach to reproducibly produce 3D spheroids or organoids that are stable for long periods of time are clinostat 'microgravity' cultures in micro-bioreactors. In these spheroids, the majority of cells experience hypoxia and glucose starvation. These conditions are certainly closer to those present in tissues than those experienced by cells in classical 2D cultures (which typically experience hyperoxia and hyperglycaemia), and are therefore critical to take into account when designing a micro-bioreactor. The recovery of physiological behavior stems from:

1. Oxygen limitations (and to a less extent glucose) induce metabolic reprogramming from oxidative phosphorylation to aerobic glycolysis and result in a strong anabolic phenotype.
2. The metabolic reprogramming includes an activation of glutaminolysis (via extra-mitochondrial pathways) (consistent with physiological increases in lipid and cholesterol synthesis).
3. Glutamine conversion to the lipid 'precursor' glutamate is linked to the hexosamine pathway activation. This correlates to increased glycogen production and protein glycosylation.

4. The additional NADPH needed for citrate and lipid synthesis is mainly generated by pentose phosphate pathway activation. Increases in acetyl-CoA also provide precursors for the observed histone acetylation.

5. Signalling pathway activities (activation of mTOR and p53, repression of NF-κB and canonical Wnt) are consistent with significant retardation of proliferation and the accumulation of cells in G1/G0, (resulting in a rate resembling that seen in both healthy and transformed cells in tissues and tumours).

6. The reduction in proliferation rate allows the cell to achieve higher ATP levels.

7. Activation of the non-canonical Wnt signalling pathway orchestrates the significant ultrastructural changes.

8. The rate of proliferation is not coupled to aerobic glycolysis.

9. Metabolic reprogramming underpins the recovery of traits mimicking in vivo physiology.

3D tissues offer an exciting model to investigate in vivo-like functionality where cells are grown in conditions that are not drastically different to those seen in vivo. Given the right growth conditions, cells 'spontaneously' revert to an in vivo mimetic physiological performance.

Supplementary Materials: The following are available online at www.mdpi.com/2306-5354/5/1/22/s1, Video S1: 17 day old spheroids in clinostat culture.

Acknowledgments: This work was supported in part by a grant from MC2 Therapeutics, Hørsholm, Denmark and from the University of Southern Denmark. We would also like to acknowledge the support of the COST actions CM1407 (Challenging organic synthesis inspired by nature—from natural products chemistry to drug discovery) and CA16119 (In vitro 3-D total cell guidance and fitness). The funding sponsors had no role in the design of the study; in the collection, analyses, or interpretation of data; in the writing of the manuscript, and in the decision to publish the results. We would like to thank Kira Eyd Joensen for excellent technical assistance, Aleksandra Amaladas for help with preparation of glucose/glycogen utilization experiment and Adelina Rogowska-Wrzesinska for critical review of the manuscript.

Author Contributions: K.W. and S.J.F. conceived and designed the experiments, analyzed the data and wrote the paper.

Conflicts of Interest: K.W. and S.J.F. are owners of CelVivo IVS, a company producing equipment and micro-bioreactors for 3D cell culture.

References

1. Balasubramanian, S.; Packard, J.A.; Leach, J.B.; Powell, E.M. Three-dimensional environment sustains morphological heterogeneity and promotes phenotypic progression during astrocyte development. *Tissue Eng. Part A* **2016**, *22*, 885–898. [CrossRef] [PubMed]

2. Drost, J.; Karthaus, W.R.; Gao, D.; Driehuis, E.; Sawyers, C.L.; Chen, Y.; Clevers, H. Organoid culture systems for prostate epithelial and cancer tissue. *Nat. Protoc.* **2016**, *11*, 347–358. [CrossRef] [PubMed]

3. Bersini, S.; Moretti, M. 3D functional and perfusable microvascular networks for organotypic microfluidic models. *J. Mater. Sci. Mater. Med.* **2015**, *26*, 180. [CrossRef] [PubMed]

4. Ashley, N.; Jones, M.; Ouaret, D.; Wilding, J.; Bodmer, W.F. Rapidly derived colorectal cancer cultures recapitulate parental cancer characteristics and enable personalized therapeutic assays. *J. Pathol.* **2014**, *234*, 34–45. [CrossRef] [PubMed]

5. Lee, S.H.; Hong, J.H.; Park, H.K.; Park, J.S.; Kim, B.K.; Lee, J.Y.; Jeong, J.Y.; Yoon, G.S.; Inoue, M.; Choi, G.S.; et al. Colorectal cancer-derived tumor spheroids retain the characteristics of original tumors. *Cancer Lett.* **2015**, *367*, 34–42. [CrossRef] [PubMed]

6. Ruppen, J.; Wildhaber, F.D.; Strub, C.; Hall, S.R.; Schmid, R.A.; Geiser, T.; Guenat, O.T. Towards personalized medicine: Chemosensitivity assays of patient lung cancer cell spheroids in a perfused microfluidic platform. *Lab Chip* **2015**, *15*, 3076–3085. [CrossRef] [PubMed]

7. Rajcevic, U.; Knol, J.C.; Piersma, S.; Bougnaud, S.; Fack, F.; Sundlisaeter, E.; Sondenaa, K.; Myklebust, R.; Pham, T.V.; Niclou, S.P.; et al. Colorectal cancer derived organotypic spheroids maintain essential tissue characteristics but adapt their metabolism in culture. *Proteome Sci.* **2014**, *12*, 39. [CrossRef] [PubMed]

8. Horning, J.L.; Sahoo, S.K.; Vijayaraghavalu, S.; Dimitrijevic, S.; Vasir, J.K.; Jain, T.K.; Panda, A.K.; Labhasetwar, V. 3-D tumor model for in vitro evaluation of anticancer drugs. *Mol. Pharm.* **2008**, *5*, 849–862. [CrossRef] [PubMed]

9. Vantangoli, M.M.; Wilson, S.; Madnick, S.J.; Huse, S.M.; Boekelheide, K. Morphologic effects of estrogen stimulation on 3D MCF-7 microtissues. *Toxicol. Lett.* **2016**, *248*, 1–8. [CrossRef] [PubMed]

10. Samuelson, L.; Gerber, D.A. Improved function and growth of pancreatic cells in a three-dimensional bioreactor environment. *Tissue Eng. Part C Methods* **2013**, *19*, 39–47. [CrossRef] [PubMed]

11. Joo, D.J.; Kim, J.Y.; Lee, J.I.; Jeong, J.H.; Cho, Y.; Ju, M.K.; Huh, K.H.; Kim, M.S.; Kim, Y.S. Manufacturing of insulin-secreting spheroids with the RIN-5F cell line using a shaking culture method. *Transplant. Proc.* **2010**, *42*, 4225–4227. [CrossRef] [PubMed]

12. Morabito, C.; Steimberg, N.; Mazzoleni, G.; Guarnieri, S.; Fano-Illic, G.; Mariggio, M.A. RCCS bioreactor-based modelled microgravity induces significant changes on in vitro 3D neuroglial cell cultures. *BioMed Res. Int.* **2015**, *2015*, 754283. [CrossRef] [PubMed]

13. Loessner, D.; Stok, K.S.; Lutolf, M.P.; Hutmacher, D.W.; Clements, J.A.; Rizzi, S.C. Bioengineered 3D platform to explore cell-ECM interactions and drug resistance of epithelial ovarian cancer cells. *Biomaterials* **2010**, *31*, 8494–8506. [CrossRef] [PubMed]

14. Grummer, R.; Hohn, H.P.; Mareel, M.M.; Denker, H.W. Adhesion and invasion of three human choriocarcinoma cell lines into human endometrium in a three-dimensional organ culture system. *Placenta* **1994**, *15*, 411–429. [CrossRef]

15. Ramaiahgari, S.C.; den Braver, M.W.; Herpers, B.; Terpstra, V.; Commandeur, J.N.; van de Water, B.; Price, L.S. A 3D in vitro model of differentiated HepG2 cell spheroids with improved liver-like properties for repeated dose high-throughput toxicity studies. *Arch. Toxicol.* **2014**, *88*, 1083–1095. [CrossRef] [PubMed]

16. Kosaka, T.; Tsuboi, S.; Fukaya, K.; Pu, H.; Ohno, T.; Tsuji, T.; Miyazaki, M.; Namba, M. Spheroid cultures of human hepatoblastoma cells (HuH-6 line) and their application for cytotoxicity assay of alcohols. *Acta Med. Okayama* **1996**, *50*, 61–66. [PubMed]

17. Fey, S.J.; Wrzesinski, K. Determination of acute lethal and chronic lethal dose thresholds of valproic acid using 3D spheroids constructed from the immortal human hepatocyte cell line HepG2/C3A. In *Valproic Acid*; Boucher, A., Ed.; Nova Science Publishers, Inc.: New York, NY, USA, 2013; pp. 141–165.

18. Wrzesinski, K.; Fey, S.J. After trypsinisation, 3D spheroids of C3A hepatocytes need 18 days to re-establish similar levels of key physiological functions to those seen in the liver. *Toxicol. Res.* **2013**, *2*, 123–135. [CrossRef]

19. Wrzesinski, K.; Magnone, M.C.; Visby Hansen, L.; Kruse, M.E.; Bergauer, T.; Bobadilla, M.; Gubler, M.; Mizrahi, J.; Zhang, K.; Andreasen, C.M.; et al. HepG2/C3a 3D spheroids exhibit stable physiological functionality for at least 24 days after recovering from trypsinisation. *Toxicol. Res.* **2013**, *2*, 163–172. [CrossRef]

20. Fatehullah, A.; Tan, S.H.; Barker, N. Organoids as an in vitro model of human development and disease. *Nat. Cell Biol.* **2016**, *18*, 246–254. [CrossRef] [PubMed]

21. Clevers, H. Modeling development and disease with organoids. *Cell* **2016**, *165*, 1586–1597. [CrossRef] [PubMed]

22. Nakano, T.; Ando, S.; Takata, N.; Kawada, M.; Muguruma, K.; Sekiguchi, K.; Saito, K.; Yonemura, S.; Eiraku, M.; Sasai, Y. Self-formation of optic cups and storable stratified neural retina from human ESCS. *Cell Stem Cell* **2012**, *10*, 771–785. [CrossRef] [PubMed]

23. Greggio, C.; De Franceschi, F.; Figueiredo-Larsen, M.; Gobaa, S.; Ranga, A.; Semb, H.; Lutolf, M.; Grapin-Botton, A. Artificial three-dimensional niches deconstruct pancreas development in vitro. *Development* **2013**, *140*, 4452–4462. [CrossRef] [PubMed]

24. McCracken, K.W.; Cata, E.M.; Crawford, C.M.; Sinagoga, K.L.; Schumacher, M.; Rockich, B.E.; Tsai, Y.H.; Mayhew, C.N.; Spence, J.R.; Zavros, Y.; et al. Modelling human development and disease in pluripotent stem-cell-derived gastric organoids. *Nature* **2014**, *516*, 400–404. [CrossRef] [PubMed]

25. Caiazzo, M.; Okawa, Y.; Ranga, A.; Piersigilli, A.; Tabata, Y.; Lutolf, M.P. Defined three-dimensional microenvironments boost induction of pluripotency. *Nat. Mater.* **2016**, *15*, 344–352. [CrossRef] [PubMed]

26. Peloso, A.; Dhal, A.; Zambon, J.P.; Li, P.; Orlando, G.; Atala, A.; Soker, S. Current achievements and future perspectives in whole-organ bioengineering. *Stem Cell Res. Ther.* **2015**, *6*, 107. [CrossRef] [PubMed]

27. Croughan, M.S.; Wang, D.I. Hydrodynamic effects on animal cells in microcarrier bioreactors. *Biotechnology* **1991**, *17*, 213–249. [PubMed]

28. Cinbiz, M.N.; Tigli, R.S.; Beskardes, I.G.; Gumusderelioglu, M.; Colak, U. Computational fluid dynamics modeling of momentum transport in rotating wall perfused bioreactor for cartilage tissue engineering. *J. Biotechnol.* **2010**, *150*, 389–395. [CrossRef] [PubMed]
29. Tsai, A.C.; Liu, Y.; Yuan, X.; Chella, R.; Ma, T. Aggregation kinetics of human mesenchymal stem cells under wave motion. *Biotechnol. J.* **2017**, *12*, 1600448. [CrossRef] [PubMed]
30. Kalmbach, A.; Bordas, R.; Oncul, A.A.; Thevenin, D.; Genzel, Y.; Reichl, U. Experimental characterization of flow conditions in 2- and 20-L bioreactors with wave-induced motion. *Biotechnol. Prog.* **2011**, *27*, 402–409. [CrossRef] [PubMed]
31. Jang, K.J.; Mehr, A.P.; Hamilton, G.A.; McPartlin, L.A.; Chung, S.; Suh, K.Y.; Ingber, D.E. Human kidney proximal tubule-on-a-chip for drug transport and nephrotoxicity assessment. *Integr. Biol.* **2013**, *5*, 1119–1129. [CrossRef] [PubMed]
32. Esch, M.B.; Prot, J.M.; Wang, Y.I.; Miller, P.; Llamas-Vidales, J.R.; Naughton, B.A.; Applegate, D.R.; Shuler, M.L. Multi-cellular 3D human primary liver cell culture elevates metabolic activity under fluidic flow. *Lab Chip* **2015**, *15*, 2269–2277. [CrossRef] [PubMed]
33. Sousa, M.F.; Silva, M.M.; Giroux, D.; Hashimura, Y.; Wesselschmidt, R.; Lee, B.; Roldao, A.; Carrondo, M.J.; Alves, P.M.; Serra, M. Production of oncolytic adenovirus and human mesenchymal stem cells in a single-use, vertical-wheel bioreactor system: Impact of bioreactor design on performance of microcarrier-based cell culture processes. *Biotechnol. Prog.* **2015**, *31*, 1600–1612. [CrossRef] [PubMed]
34. Ismadi, M.Z.; Gupta, P.; Fouras, A.; Verma, P.; Jadhav, S.; Bellare, J.; Hourigan, K. Flow characterization of a spinner flask for induced pluripotent stem cell culture application. *PLoS ONE* **2014**, *9*, e106493. [CrossRef] [PubMed]
35. Dardik, A.; Chen, L.; Frattini, J.; Asada, H.; Aziz, F.; Kudo, F.A.; Sumpio, B.E. Differential effects of orbital and laminar shear stress on endothelial cells. *J. Vasc. Surg.* **2005**, *41*, 869–880. [CrossRef] [PubMed]
36. Gareau, T.; Lara, G.G.; Shepherd, R.D.; Krawetz, R.; Rancourt, D.E.; Rinker, K.D.; Kallos, M.S. Shear stress influences the pluripotency of murine embryonic stem cells in stirred suspension bioreactors. *J. Tissue Eng. Regen. Med.* **2014**, *8*, 268–278. [CrossRef] [PubMed]
37. Bai, G.; Bee, J.S.; Biddlecombe, J.G.; Chen, Q.; Leach, W.T. Computational fluid dynamics (CFD) insights into agitation stress methods in biopharmaceutical development. *Int. J. Pharm.* **2012**, *423*, 264–280. [CrossRef] [PubMed]
38. Filipovic, N.; Ghimire, K.; Saveljic, I.; Milosevic, Z.; Ruegg, C. Computational modeling of shear forces and experimental validation of endothelial cell responses in an orbital well shaker system. *Comput. Methods Biomech. Biomed. Eng.* **2016**, *19*, 581–590. [CrossRef] [PubMed]
39. Warburg, O. On the origin of cancer cells. *Science* **1956**, *123*, 309–314. [CrossRef] [PubMed]
40. Moreno-Sanchez, R.; Rodriguez-Enriquez, S.; Marin-Hernandez, A.; Saavedra, E. Energy metabolism in tumor cells. *FEBS J.* **2007**, *274*, 1393–1418. [CrossRef] [PubMed]
41. Vander Heiden, M.G.; Cantley, L.C.; Thompson, C.B. Understanding the warburg effect: The metabolic requirements of cell proliferation. *Science* **2009**, *324*, 1029–1033. [CrossRef] [PubMed]
42. Day, S.E.; Kettunen, M.I.; Gallagher, F.A.; Hu, D.E.; Lerche, M.; Wolber, J.; Golman, K.; Ardenkjaer-Larsen, J.H.; Brindle, K.M. Detecting tumor response to treatment using hyperpolarized ^{13}C magnetic resonance imaging and spectroscopy. *Nat. Med.* **2007**, *13*, 1382–1387. [CrossRef] [PubMed]
43. Albers, M.J.; Bok, R.; Chen, A.P.; Cunningham, C.H.; Zierhut, M.L.; Zhang, V.Y.; Kohler, S.J.; Tropp, J.; Hurd, R.E.; Yen, Y.F.; et al. Hyperpolarized ^{13}C lactate, pyruvate, and alanine: Noninvasive biomarkers for prostate cancer detection and grading. *Cancer Res.* **2008**, *68*, 8607–8615. [CrossRef] [PubMed]
44. Rodrigues, T.B.; Serrao, E.M.; Kennedy, B.W.; Hu, D.E.; Kettunen, M.I.; Brindle, K.M. Magnetic resonance imaging of tumor glycolysis using hyperpolarized ^{13}C-labeled glucose. *Nat. Med.* **2014**, *20*, 93–97. [CrossRef] [PubMed]
45. Shannon, B.J.; Vaishnavi, S.N.; Vlassenko, A.G.; Shimony, J.S.; Rutlin, J.; Raichle, M.E. Brain aerobic glycolysis and motor adaptation learning. *Proc. Natl. Acad. Sci. USA* **2016**, *113*, E3782–E3791. [CrossRef] [PubMed]
46. Lumata, L.; Yang, C.; Ragavan, M.; Carpenter, N.; DeBerardinis, R.J.; Merritt, M.E. Hyperpolarized (13)C magnetic resonance and its use in metabolic assessment of cultured cells and perfused organs. *Methods Enzymol.* **2015**, *561*, 73–106. [PubMed]

47. Fan, T.W.; Kucia, M.; Jankowski, K.; Higashi, R.M.; Ratajczak, J.; Ratajczak, M.Z.; Lane, A.N. Rhabdomyosarcoma cells show an energy producing anabolic metabolic phenotype compared with primary myocytes. *Mol. Cancer* **2008**, *7*, 79. [CrossRef] [PubMed]

48. Fantin, V.R.; St-Pierre, J.; Leder, P. Attenuation of LDH-A expression uncovers a link between glycolysis, mitochondrial physiology, and tumor maintenance. *Cancer Cell* **2006**, *9*, 425–434. [CrossRef] [PubMed]

49. Hatzivassiliou, G.; Zhao, F.; Bauer, D.E.; Andreadis, C.; Shaw, A.N.; Dhanak, D.; Hingorani, S.R.; Tuveson, D.A.; Thompson, C.B. ATP citrate lyase inhibition can suppress tumor cell growth. *Cancer Cell* **2005**, *8*, 311–321. [CrossRef] [PubMed]

50. Weinberg, F.; Hamanaka, R.; Wheaton, W.W.; Weinberg, S.; Joseph, J.; Lopez, M.; Kalyanaraman, B.; Mutlu, G.M.; Budinger, G.R.; Chandel, N.S. Mitochondrial metabolism and ROS generation are essential for Kras-mediated tumorigenicity. *Proc. Natl. Acad. Sci. USA* **2010**, *107*, 8788–8793. [CrossRef] [PubMed]

51. Nakajima, T.; Moriguchi, M.; Mitsumoto, Y.; Katagishi, T.; Kimura, H.; Shintani, H.; Deguchi, T.; Okanoue, T.; Kagawa, K.; Ashihara, T. Simple tumor profile chart based on cell kinetic parameters and histologic grade is useful for estimating the natural growth rate of hepatocellular carcinoma. *Hum. Pathol.* **2002**, *33*, 92–99. [CrossRef] [PubMed]

52. Zu, X.L.; Guppy, M. Cancer metabolism: Facts, fantasy, and fiction. *Biochem. Biophys. Res. Commun.* **2004**, *313*, 459–465. [CrossRef] [PubMed]

53. Ward, P.S.; Thompson, C.B. Metabolic reprogramming: A cancer hallmark even warburg did not anticipate. *Cancer Cell* **2012**, *21*, 297–308. [CrossRef] [PubMed]

54. DeBerardinis, R.J.; Lum, J.J.; Hatzivassiliou, G.; Thompson, C.B. The biology of cancer: Metabolic reprogramming fuels cell growth and proliferation. *Cell Metab.* **2008**, *7*, 11–20. [CrossRef] [PubMed]

55. Klein, C.A. Selection and adaptation during metastatic cancer progression. *Nature* **2013**, *501*, 365–372. [CrossRef] [PubMed]

56. Sabo, A.; Kress, T.R.; Pelizzola, M.; de Pretis, S.; Gorski, M.M.; Tesi, A.; Morelli, M.J.; Bora, P.; Doni, M.; Verrecchia, A.; et al. Selective transcriptional regulation by Myc in cellular growth control and lymphomagenesis. *Nature* **2014**, *511*, 488–492. [CrossRef] [PubMed]

57. Wrzesinski, K.; Rogowska-Wrzesinska, A.; Kanlaya, R.; Borkowski, K.; Schwammle, V.; Dai, J.; Joensen, K.E.; Wojdyla, K.; Carvalho, V.B.; Fey, S.J. The cultural divide: Exponential growth in classical 2d and metabolic equilibrium in 3D environments. *PLoS ONE* **2014**, *9*, e106973. [CrossRef] [PubMed]

58. Jiang, L.; Shestov, A.A.; Swain, P.; Yang, C.; Parker, S.J.; Wang, Q.A.; Terada, L.S.; Adams, N.D.; McCabe, M.T.; Pietrak, B.; et al. Reductive carboxylation supports redox homeostasis during anchorage-independent growth. *Nature* **2016**, *532*, 255–258. [CrossRef] [PubMed]

59. Wrzesinski, K.; Fey, S.J. From 2D to 3D—A new dimension for modelling the effect of natural products on human tissue. *Curr. Pharm. Des.* **2015**, *21*, 5605–5616. [CrossRef] [PubMed]

60. Rogowska-Wrzesinska, A.; Wrzesinski, K.; Fey, S.J. Heteromer score-using internal standards to assess the quality of proteomic data. *Proteomics* **2014**, *14*, 1042–1047. [CrossRef] [PubMed]

61. Metallo, C.M.; Gameiro, P.A.; Bell, E.L.; Mattaini, K.R.; Yang, J.; Hiller, K.; Jewell, C.M.; Johnson, Z.R.; Irvine, D.J.; Guarente, L.; et al. Reductive glutamine metabolism by IDH1 mediates lipogenesis under hypoxia. *Nature* **2012**, *481*, 380–384. [CrossRef] [PubMed]

62. Mazurek, S.; Boschek, C.B.; Hugo, F.; Eigenbrodt, E. Pyruvate kinase type M2 and its role in tumor growth and spreading. *Semin. Cancer Biol.* **2005**, *15*, 300–308. [CrossRef] [PubMed]

63. Christofk, H.R.; Vander Heiden, M.G.; Harris, M.H.; Ramanathan, A.; Gerszten, R.E.; Wei, R.; Fleming, M.D.; Schreiber, S.L.; Cantley, L.C. The M2 splice isoform of pyruvate kinase is important for cancer metabolism and tumour growth. *Nature* **2008**, *452*, 230–233. [CrossRef] [PubMed]

64. Chaneton, B.; Gottlieb, E. Rocking cell metabolism: Revised functions of the key glycolytic regulator PKM2 in cancer. *Trends Biochem. Sci.* **2012**, *37*, 309–316. [CrossRef] [PubMed]

65. Zaidi, N.; Swinnen, J.V.; Smans, K. ATP-citrate lyase: A key player in cancer metabolism. *Cancer Res.* **2012**, *72*, 3709–3714. [CrossRef] [PubMed]

66. Wang, G.; Sai, K.; Gong, F.; Yang, Q.; Chen, F.; Lin, J. Mutation of isocitrate dehydrogenase 1 induces glioma cell proliferation via nuclear factor-kappaB activation in a hypoxia-inducible factor 1-alpha dependent manner. *Mol. Med. Rep.* **2014**, *9*, 1799–1805. [CrossRef] [PubMed]

67. Buck, M.D.; O'Sullivan, D.; Klein Geltink, R.I.; Curtis, J.D.; Chang, C.H.; Sanin, D.E.; Qiu, J.; Kretz, O.; Braas, D.; van der Windt, G.J.; et al. Mitochondrial dynamics controls T cell fate through metabolic programming. *Cell* **2016**, *166*, 63–76. [CrossRef] [PubMed]

68. Yang, C.; Harrison, C.; Jin, E.S.; Chuang, D.T.; Sherry, A.D.; Malloy, C.R.; Merritt, M.E.; DeBerardinis, R.J. Simultaneous steady-state and dynamic ^{13}C NMR can differentiate alternative routes of pyruvate metabolism in living cancer cells. *J. Biol. Chem.* **2014**, *289*, 6212–6224. [CrossRef] [PubMed]

69. Keshari, K.R.; Kurhanewicz, J.; Jeffries, R.E.; Wilson, D.M.; Dewar, B.J.; Van Criekinge, M.; Zierhut, M.; Vigneron, D.B.; Macdonald, J.M. Hyperpolarized (13)C spectroscopy and an NMR-compatible bioreactor system for the investigation of real-time cellular metabolism. *Magn. Reson. Med.* **2010**, *63*, 322–329. [CrossRef] [PubMed]

70. DeBerardinis, R.J.; Mancuso, A.; Daikhin, E.; Nissim, I.; Yudkoff, M.; Wehrli, S.; Thompson, C.B. Beyond aerobic glycolysis: Transformed cells can engage in glutamine metabolism that exceeds the requirement for protein and nucleotide synthesis. *Proc. Natl. Acad. Sci. USA* **2007**, *104*, 19345–19350. [CrossRef] [PubMed]

71. Carmeliet, P.; Jain, R.K. Angiogenesis in cancer and other diseases. *Nature* **2000**, *407*, 249–257. [CrossRef] [PubMed]

72. Grimes, D.R.; Kelly, C.; Bloch, K.; Partridge, M. A method for estimating the oxygen consumption rate in multicellular tumour spheroids. *J. R. Soc. Interface* **2014**, *11*, 20131124. [CrossRef] [PubMed]

73. Carreau, A.; El Hafny-Rahbi, B.; Matejuk, A.; Grillon, C.; Kieda, C. Why is the partial oxygen pressure of human tissues a crucial parameter? Small molecules and hypoxia. *J. Cell. Mol. Med.* **2011**, *15*, 1239–1253. [CrossRef] [PubMed]

74. Lenihan, C.R.; Taylor, C.T. The impact of hypoxia on cell death pathways. *Biochem. Soc. Trans.* **2013**, *41*, 657–663. [CrossRef] [PubMed]

75. Papandreou, I.; Krishna, C.; Kaper, F.; Cai, D.; Giaccia, A.J.; Denko, N.C. Anoxia is necessary for tumor cell toxicity caused by a low-oxygen environment. *Cancer Res.* **2005**, *65*, 3171–3178. [CrossRef] [PubMed]

76. Dmitriev, R.I.; Zhdanov, A.V.; Nolan, Y.M.; Papkovsky, D.B. Imaging of neurosphere oxygenation with phosphorescent probes. *Biomaterials* **2013**, *34*, 9307–9317. [CrossRef] [PubMed]

77. Sutherland, R.M. Cell and environment interactions in tumor microregions: The multicell spheroid model. *Science* **1988**, *240*, 177–184. [CrossRef] [PubMed]

78. Mueller-Klieser, W.F.; Sutherland, R.M. Influence of convection in the growth medium on oxygen tensions in multicellular tumor spheroids. *Cancer Res.* **1982**, *42*, 237–242. [PubMed]

79. Hulikova, A.; Swietach, P. Rapid CO_2 permeation across biological membranes: Implications for CO_2 venting from tissue. *FASEB J.* **2014**, *28*, 2762–2774. [CrossRef] [PubMed]

80. Hulikova, A.; Vaughan-Jones, R.D.; Swietach, P. Dual role of $CO_2/HCO_3(-)$ buffer in the regulation of intracellular pH of three-dimensional tumor growths. *J. Biol. Chem.* **2011**, *286*, 13815–13826. [CrossRef] [PubMed]

81. Sato, M.; Kawana, K.; Adachi, K.; Fujimoto, A.; Yoshida, M.; Nakamura, H.; Nishida, H.; Inoue, T.; Taguchi, A.; Takahashi, J.; et al. Spheroid cancer stem cells display reprogrammed metabolism and obtain energy by actively running the tricarboxylic acid (TCA) cycle. *Oncotarget* **2016**, *7*, 33297–33305. [CrossRef] [PubMed]

82. Fukuda, R.; Zhang, H.; Kim, J.W.; Shimoda, L.; Dang, C.V.; Semenza, G.L. HIF-1 regulates cytochrome oxidase subunits to optimize efficiency of respiration in hypoxic cells. *Cell* **2007**, *129*, 111–122. [CrossRef] [PubMed]

83. Chen, K.; Chen, S.; Huang, C.; Cheng, H.; Zhou, R. TCTP increases stability of hypoxia-inducible factor 1alpha by interaction with and degradation of the tumour suppressor VHL. *Biol. Cell* **2013**, *105*, 208–218. [CrossRef] [PubMed]

84. Lum, J.J.; Bui, T.; Gruber, M.; Gordan, J.D.; DeBerardinis, R.J.; Covello, K.L.; Simon, M.C.; Thompson, C.B. The transcription factor HIF-1alpha plays a critical role in the growth factor-dependent regulation of both aerobic and anaerobic glycolysis. *Genes Dev.* **2007**, *21*, 1037–1049. [CrossRef] [PubMed]

85. Gilkes, D.M.; Bajpai, S.; Chaturvedi, P.; Wirtz, D.; Semenza, G.L. Hypoxia-inducible factor 1 (HIF-1) promotes extracellular matrix remodeling under hypoxic conditions by inducing P4HA1, P4HA2, and PLOD2 expression in fibroblasts. *J. Biol. Chem.* **2013**, *288*, 10819–10829. [CrossRef] [PubMed]

86. Quiros, P.M.; Espanol, Y.; Acin-Perez, R.; Rodriguez, F.; Barcena, C.; Watanabe, K.; Calvo, E.; Loureiro, M.; Fernandez-Garcia, M.S.; Fueyo, A.; et al. ATP-dependent lon protease controls tumor bioenergetics by reprogramming mitochondrial activity. *Cell Rep.* **2014**, *8*, 542–556. [CrossRef] [PubMed]

87. Wojdyla, K.; Wrzesinski, K.; Williamson, J.; Fey, S.J.; Rogowska-Wrzesinska, A. Acetaminophen-induced *S*-nitrosylation and *S*-sulfenylation changes in 3D cultured hepatocarcinoma cell spheroids. *Toxicol. Res.* **2016**, *5*, 905–920. [CrossRef]

88. Lee, A.S. Glucose-regulated proteins in cancer: Molecular mechanisms and therapeutic potential. *Nat. Rev. Cancer* **2014**, *14*, 263–276. [CrossRef] [PubMed]

89. Bravo, R.; Parra, V.; Gatica, D.; Rodriguez, A.E.; Torrealba, N.; Paredes, F.; Wang, Z.V.; Zorzano, A.; Hill, J.A.; Jaimovich, E.; et al. Endoplasmic reticulum and the unfolded protein response: Dynamics and metabolic integration. *Int. Rev. Cell Mol. Biol.* **2013**, *301*, 215–290. [PubMed]

90. Korennykh, A.; Walter, P. Structural basis of the unfolded protein response. *Annu. Rev. Cell Dev. Biol.* **2012**, *28*, 251–277. [CrossRef] [PubMed]

91. Behnke, J.; Feige, M.J.; Hendershot, L.M. BiP and its nucleotide exchange factors Grp170 and Sil1: Mechanisms of action and biological functions. *J. Mol. Biol.* **2015**, *427*, 1589–1608. [CrossRef] [PubMed]

92. Bhutia, Y.D.; Ganapathy, V. Glutamine transporters in mammalian cells and their functions in physiology and cancer. *Biochim. Biophys. Acta* **2016**, *1863*, 2531–2539. [CrossRef] [PubMed]

93. Jeon, Y.J.; Khelifa, S.; Ratnikov, B.; Scott, D.A.; Feng, Y.; Parisi, F.; Ruller, C.; Lau, E.; Kim, H.; Brill, L.M.; et al. Regulation of glutamine carrier proteins by RNF5 determines breast cancer response to ER stress-inducing chemotherapies. *Cancer Cell* **2015**, *27*, 354–369. [CrossRef] [PubMed]

94. Broer, A.; Rahimi, F.; Broer, S. Deletion of amino acid transporter ASCT2 (SLC1A5) reveals an essential role for transporters SNAT1 (SLC38A1) and SNAT2 (SLC38A2) to sustain glutaminolysis in cancer cells. *J. Biol. Chem.* **2016**, *291*, 13194–13205. [CrossRef] [PubMed]

95. Guillaumond, F.; Leca, J.; Olivares, O.; Lavaut, M.N.; Vidal, N.; Berthezene, P.; Dusetti, N.J.; Loncle, C.; Calvo, E.; Turrini, O.; et al. Strengthened glycolysis under hypoxia supports tumor symbiosis and hexosamine biosynthesis in pancreatic adenocarcinoma. *Proc. Natl. Acad. Sci. USA* **2013**, *110*, 3919–3924. [CrossRef] [PubMed]

96. Salminen, A.; Kaarniranta, K.; Kauppinen, A. Hypoxia-inducible histone lysine demethylases: Impact on the aging process and age-related diseases. *Aging Dis.* **2016**, *7*, 180–200. [PubMed]

97. Tvardovskiy, A.; Schwammle, V.; Kempf, S.J.; Rogowska-Wrzesinska, A.; Jensen, O.N. Accumulation of histone variant $H_{3.3}$ with age is associated with profound changes in the histone methylation landscape. *Nucleic Acids Res.* **2017**, *45*, 9272–9289. [CrossRef] [PubMed]

98. Lim, S.K.; Jeong, Y.W.; Kim, D.I.; Park, M.J.; Choi, J.H.; Kim, S.U.; Kang, S.S.; Han, H.J.; Park, S.H. Activation of PRMT1 and PRMT5 mediates hypoxia- and ischemia-induced apoptosis in human lung epithelial cells and the lung of miniature pigs: The role of p38 and JNK mitogen-activated protein kinases. *Biochem. Biophys. Res. Commun.* **2013**, *440*, 707–713. [CrossRef] [PubMed]

99. Reintjes, A.; Fuchs, J.E.; Kremser, L.; Lindner, H.H.; Liedl, K.R.; Huber, L.A.; Valovka, T. Asymmetric arginine dimethylation of RelA provides a repressive mark to modulate TNFalpha/NF-kappaB response. *Proc. Natl. Acad. Sci. USA* **2016**, *113*, 4326–4331. [CrossRef] [PubMed]

100. Takai, H.; Masuda, K.; Sato, T.; Sakaguchi, Y.; Suzuki, T.; Suzuki, T.; Koyama-Nasu, R.; Nasu-Nishimura, Y.; Katou, Y.; Ogawa, H.; et al. 5-hydroxymethylcytosine plays a critical role in glioblastomagenesis by recruiting the CHTOP-methylosome complex. *Cell Rep.* **2014**, *9*, 48–60. [CrossRef] [PubMed]

101. Lunt, S.Y.; Vander Heiden, M.G. Aerobic glycolysis: Meeting the metabolic requirements of cell proliferation. *Annu. Rev. Cell Dev. Biol.* **2011**, *27*, 441–464. [CrossRef] [PubMed]

102. Ng, B.G.; Wolfe, L.A.; Ichikawa, M.; Markello, T.; He, M.; Tifft, C.J.; Gahl, W.A.; Freeze, H.H. Biallelic mutations in cad, impair de novo pyrimidine biosynthesis and decrease glycosylation precursors. *Hum. Mol. Genet.* **2015**, *24*, 3050–3057. [CrossRef] [PubMed]

103. Dibble, C.C.; Cantley, L.C. Regulation of mTORC1 by PI3K signaling. *Trends Cell Biol.* **2015**, *25*, 545–555. [CrossRef] [PubMed]

104. Duvel, K.; Yecies, J.L.; Menon, S.; Raman, P.; Lipovsky, A.I.; Souza, A.L.; Triantafellow, E.; Ma, Q.; Gorski, R.; Cleaver, S.; et al. Activation of a metabolic gene regulatory network downstream of mTOR complex 1. *Mol. Cell* **2010**, *39*, 171–183. [CrossRef] [PubMed]

105. Robitaille, A.M.; Christen, S.; Shimobayashi, M.; Cornu, M.; Fava, L.L.; Moes, S.; Prescianotto-Baschong, C.; Sauer, U.; Jenoe, P.; Hall, M.N. Quantitative phosphoproteomics reveal mTORC1 activates de novo pyrimidine synthesis. *Science* **2013**, *339*, 1320–1323. [CrossRef] [PubMed]

106. Peterson, T.R.; Sengupta, S.S.; Harris, T.E.; Carmack, A.E.; Kang, S.A.; Balderas, E.; Guertin, D.A.; Madden, K.L.; Carpenter, A.E.; Finck, B.N.; et al. mTOR complex 1 regulates lipin 1 localization to control the SREBP pathway. *Cell* **2011**, *146*, 408–420. [CrossRef] [PubMed]

107. Stine, Z.E.; Walton, Z.E.; Altman, B.J.; Hsieh, A.L.; Dang, C.V. Myc, metabolism, and cancer. *Cancer Discov.* **2015**, *5*, 1024–1039. [CrossRef] [PubMed]

108. Locasale, J.W. Serine, glycine and one-carbon units: Cancer metabolism in full circle. *Nat. Rev. Cancer* **2013**, *13*, 572–583. [CrossRef] [PubMed]

109. Osthus, R.C.; Shim, H.; Kim, S.; Li, Q.; Reddy, R.; Mukherjee, M.; Xu, Y.; Wonsey, D.; Lee, L.A.; Dang, C.V. Deregulation of glucose transporter 1 and glycolytic gene expression by c-Myc. *J. Biol. Chem.* **2000**, *275*, 21797–21800. [CrossRef] [PubMed]

110. Kim, J.W.; Gao, P.; Liu, Y.C.; Semenza, G.L.; Dang, C.V. Hypoxia-inducible factor 1 and dysregulated c-Myc cooperatively induce vascular endothelial growth factor and metabolic switches hexokinase 2 and pyruvate dehydrogenase kinase 1. *Mol. Cell. Biol.* **2007**, *27*, 7381–7393. [CrossRef] [PubMed]

111. Koshiji, M.; Kageyama, Y.; Pete, E.A.; Horikawa, I.; Barrett, J.C.; Huang, L.E. HIF-1alpha induces cell cycle arrest by functionally counteracting Myc. *EMBO J.* **2004**, *23*, 1949–1956. [CrossRef] [PubMed]

112. David, C.J.; Chen, M.; Assanah, M.; Canoll, P.; Manley, J.L. HnRNP proteins controlled by c-Myc deregulate pyruvate kinase mRNA splicing in cancer. *Nature* **2010**, *463*, 364–368. [CrossRef] [PubMed]

113. Wonsey, D.R.; Zeller, K.I.; Dang, C.V. The c-Myc target gene PRDX3 is required for mitochondrial homeostasis and neoplastic transformation. *Proc. Natl. Acad. Sci. USA* **2002**, *99*, 6649–6654. [CrossRef] [PubMed]

114. Moon, H.J.; Redman, K.L. Trm4 and Nsun2 RNA:m^{5}c methyltransferases form metabolite-dependent, covalent adducts with previously methylated RNA. *Biochemistry* **2014**, *53*, 7132–7144. [CrossRef] [PubMed]

115. Yang, H.; Li, T.W.; Ko, K.S.; Xia, M.; Lu, S.C. Switch from Mnt-Max to Myc-Max induces p53 and cyclin D1 expression and apoptosis during cholestasis in mouse and human hepatocytes. *Hepatology* **2009**, *49*, 860–870. [CrossRef] [PubMed]

116. Vasseur, S.; Afzal, S.; Tomasini, R.; Guillaumond, F.; Tardivel-Lacombe, J.; Mak, T.W.; Iovanna, J.L. Consequences of DJ-1 upregulation following p53 loss and cell transformation. *Oncogene* **2012**, *31*, 664–670. [CrossRef] [PubMed]

117. Moscovitz, O.; Ben-Nissan, G.; Fainer, I.; Pollack, D.; Mizrachi, L.; Sharon, M. The Parkinson's-associated protein DJ-1 regulates the 20S proteasome. *Nat. Commun.* **2015**, *6*, 6609. [CrossRef] [PubMed]

118. Mak, G.W.; Lai, W.L.; Zhou, Y.; Li, M.; Ng, I.O.; Ching, Y.P. CDK5RAP3 is a novel repressor of p14ARF in hepatocellular carcinoma cells. *PLoS ONE* **2012**, *7*, e42210. [CrossRef] [PubMed]

119. Meng, W.; Ellsworth, B.A.; Nirschl, A.A.; McCann, P.J.; Patel, M.; Girotra, R.N.; Wu, G.; Sher, P.M.; Morrison, E.P.; Biller, S.A.; et al. Discovery of dapagliflozin: A potent, selective renal sodium-dependent glucose cotransporter 2 (SGLT2) inhibitor for the treatment of type 2 diabetes. *J. Med. Chem.* **2008**, *51*, 1145–1149. [CrossRef] [PubMed]

120. Amson, R.; Pece, S.; Marine, J.C.; Di Fiore, P.P.; Telerman, A. TPT1/TCTP-regulated pathways in phenotypic reprogramming. *Trends Cell Biol.* **2013**, *23*, 37–46. [CrossRef] [PubMed]

121. Chen, W.; Wang, H.; Tao, S.; Zheng, Y.; Wu, W.; Lian, F.; Jaramillo, M.; Fang, D.; Zhang, D.D. Tumor protein translationally controlled 1 is a p53 target gene that promotes cell survival. *Cell Cycle* **2013**, *12*, 2321–2328. [CrossRef] [PubMed]

122. Amson, R.; Pece, S.; Lespagnol, A.; Vyas, R.; Mazzarol, G.; Tosoni, D.; Colaluca, I.; Viale, G.; Rodrigues-Ferreira, S.; Wynendaele, J.; et al. Reciprocal repression between p53 and TCTP. *Nat. Med.* **2012**, *18*, 91–99. [CrossRef] [PubMed]

123. Johansson, H.; Vizlin-Hodzic, D.; Simonsson, T.; Simonsson, S. Translationally controlled tumor protein interacts with nucleophosmin during mitosis in ES cells. *Cell Cycle* **2010**, *9*, 2160–2169. [CrossRef] [PubMed]

124. Graves, J.A.; Metukuri, M.; Scott, D.; Rothermund, K.; Prochownik, E.V. Regulation of reactive oxygen species homeostasis by peroxiredoxins and c-myc. *J. Biol. Chem.* **2009**, *284*, 6520–6529. [CrossRef] [PubMed]

125. Wang, B.; Zhao, L.; Fish, M.; Logan, C.Y.; Nusse, R. Self-renewing diploid Axin2(+) cells fuel homeostatic renewal of the liver. *Nature* **2015**, *524*, 180–185. [CrossRef] [PubMed]

126. Wlodarchak, N.; Xing, Y. PP2A as a master regulator of the cell cycle. *Crit. Rev. Biochem. Mol. Biol.* **2016**, *51*, 162–184. [CrossRef] [PubMed]

127. Wang, L.; Li, H.; Zhang, Y.; Santella, R.M.; Weinstein, I.B. HINT1 inhibits beta-catenin/TCF4, USF2 and NFkappab activity in human hepatoma cells. *Int. J. Cancer* **2009**, *124*, 1526–1534. [CrossRef] [PubMed]

128. Weiske, J.; Huber, O. The histidine triad protein hint1 triggers apoptosis independent of its enzymatic activity. *J. Biol. Chem.* **2006**, *281*, 27356–27366. [CrossRef] [PubMed]

129. Wu, D.; Wu, P.; Zhao, L.; Huang, L.; Zhang, Z.; Zhao, S.; Huang, J. NF-kappaB expression and outcomes in solid tumors: A systematic review and meta-analysis. *Medicine* **2015**, *94*, e1687. [CrossRef] [PubMed]

130. Clark, C.R.; Robinson, J.Y.; Sanchez, N.S.; Townsend, T.A.; Arrieta, J.A.; Merryman, W.D.; Trykall, D.Z.; Olivey, H.E.; Hong, C.C.; Barnett, J.V. Common pathways regulate type III TGFbeta receptor-dependent cell invasion in epicardial and endocardial cells. *Cell. Signal.* **2016**, *28*, 688–698. [CrossRef] [PubMed]

131. Turtoi, A.; Blomme, A.; Debois, D.; Somja, J.; Delvaux, D.; Patsos, G.; Di Valentin, E.; Peulen, O.; Mutijima, E.N.; De Pauw, E.; et al. Organized proteomic heterogeneity in colorectal cancer liver metastases and implications for therapies. *Hepatology* **2014**, *59*, 924–934. [CrossRef] [PubMed]

132. Yang, Y.; Sun, H.; Li, X.; Ding, Q.; Wei, P.; Zhou, J. Transforming growth factor beta-induced is essential for endotoxin tolerance induced by a low dose of lipopolysaccharide in human peripheral blood mononuclear cells. *Iran. J. Allergy Asthma Immunol.* **2015**, *14*, 321–330. [PubMed]

133. Galluzzi, L.; Bravo-San Pedro, J.M.; Kepp, O.; Kroemer, G. Regulated cell death and adaptive stress responses. *Cell. Mol. Life Sci. CMLS* **2016**, *73*, 2405–2410. [CrossRef] [PubMed]

134. Vaseva, A.V.; Marchenko, N.D.; Ji, K.; Tsirka, S.E.; Holzmann, S.; Moll, U.M. P53 opens the mitochondrial permeability transition pore to trigger necrosis. *Cell* **2012**, *149*, 1536–1548. [CrossRef] [PubMed]

135. Galluzzi, L.; Bravo-San Pedro, J.M.; Kroemer, G. Organelle-specific initiation of cell death. *Nat. Cell Biol.* **2014**, *16*, 728–736. [CrossRef] [PubMed]

bioengineering

MDPI

Article

Microscale 3D Liver Bioreactor for In Vitro Hepatotoxicity Testing under Perfusion Conditions

Nora Freyer [1,*,†], Selina Greuel [1,†], Fanny Knöspel [1], Florian Gerstmann [1], Lisa Storch [1], Georg Damm [2], Daniel Seehofer [2], Jennifer Foster Harris [3], Rashi Iyer [3], Frank Schubert [4] and Katrin Zeilinger [1]

[1]   Berlin-Brandenburg Center for Regenerative Therapies (BCRT), Charité–Universitätsmedizin Berlin, 13353 Berlin, Germany; selina.greuel@charite.de (S.G.); fanny.knoespel@gmx.de (F.K.); f.gerstmann@mailbox.tu-berlin.de (F.G.); lisa.storch@gmx.net (L.S.); katrin.zeilinger@charite.de (K.Z.)
[2]   Department of Hepatobiliary Surgery and Visceral Transplantation, University of Leipzig, 04103 Leipzig, Germany; georg.damm@medizin.uni-leipzig.de (G.D.); daniel.seehofer@medizin.uni-leipzig.de (D.S.)
[3]   Los Alamos National Laboratory, Los Alamos, NM 87545, USA; jfharris@lanl.gov (J.F.H.); rashi@lanl.gov (R.I.)
[4]   StemCell Systems GmbH, Berlin 12101, Germany; frank.schubert@stemcell-systems.com
*   Correspondence: nora.freyer@charite.de; Tel.: +49-30-450-552501
†   These authors contributed equally to this work.

Received: 12 February 2018; Accepted: 12 March 2018; Published: 15 March 2018

Abstract: The accurate prediction of hepatotoxicity demands validated human in vitro models that can close the gap between preclinical animal studies and clinical trials. In this study we investigated the response of primary human liver cells to toxic drug exposure in a perfused microscale 3D liver bioreactor. The cellularized bioreactors were treated with 5, 10, or 30 mM acetaminophen (APAP) used as a reference substance. Lactate production significantly decreased upon treatment with 30 mM APAP ($p < 0.05$) and ammonia release significantly increased in bioreactors treated with 10 or 30 mM APAP ($p < 0.0001$), indicating APAP-induced dose-dependent toxicity. The release of prostaglandin E2 showed a significant increase at 30 mM APAP ($p < 0.05$), suggesting an inflammatory reaction towards enhanced cellular stress. The expression of genes involved in drug metabolism, antioxidant reactions, urea synthesis, and apoptosis was differentially influenced by APAP exposure. Histological examinations revealed that primary human liver cells in untreated control bioreactors were reorganized in tissue-like cell aggregates. These aggregates were partly disintegrated upon APAP treatment, lacking expression of hepatocyte-specific proteins and transporters. In conclusion, our results validate the suitability of the microscale 3D liver bioreactor to detect hepatotoxic effects of drugs in vitro under perfusion conditions.

Keywords: microscale 3D liver bioreactor; in vitro perfusion; primary human liver cells; hepatotoxicity; acetaminophen

1. Introduction

The evaluation of hepatotoxicity of pharmaceutical substances is one major aspect of drug development. For in vivo hepatotoxicity testing, animal models, especially rats and mice, are currently the method of choice [1,2]. However, the use of such models is controversial as they often do not accurately represent the human metabolism due to differences in pharmacokinetics, pharmacodynamics, and species-specific genetic variations [3,4]. The accurate prediction of potential hepatotoxicity demands validated human in vitro models that can close the gap between preclinical animal studies and clinical trials in drug toxicity testing.

Currently applied in vitro toxicity testing models, especially in earlier developmental stages, are mostly based on 2D cell cultures, which offer several advantages including low costs, high throughput, and reproducibility. However, conventional 2D models using primary human hepatocytes are impeded by a rapid decrease of hepatic function and by cell dedifferentiation [5]. This phenomenon is partly due to loss of the original 3D architecture of the organ, which is characterized by organ-specific cell–cell and cell–extracellular matrix contacts. A promising approach to create a physiologically relevant surrounding in vitro can be seen in the development of 3D culture models [6]. Evidence shows that 3D models better reflect the microcellular environment than 2D cultures and thereby enable a more realistic prediction of in vivo drug effects [7–9]. Moreover, an even closer approximation of the in vivo situation is provided by perfused culture platforms that mimic the in vivo hemodynamics and enhance nutrient supply of the cells [10,11]. In addition, perfused culture models enable a constant exposure to test compounds with simultaneous removal of metabolites, in contrast to static 2D cultures with discontinuous medium exchange. In this context, microfluidic culture systems gain increasing importance, since they allow minimization of the amounts of cells and reagents needed while providing characteristics of 3D cultures with physiological cell arrangement [12,13].

We have previously shown that a scalable 3D multicompartment bioreactor technology with counter-current medium exchange and decentralized oxygenation supports the reorganization and longevity of primary human liver cells in vitro [14]. It was also demonstrated that the technology is suitable for analysis of hepatic drug metabolism and hepatic toxicity using serum-free conditions [15–17]. Based on the existing technology, a microscale 3D liver bioreactor with a cell compartment volume of 100 μL was constructed for applications in preclinical drug development and toxicity testing.

The goal of the present study was to investigate the potential of the device to detect toxic drug effects, using acetaminophen (APAP) as a reference substance. For this purpose, primary human liver cells were cultured in microscale 3D bioreactors over six days and treated with APAP at final concentrations of 0, 5, 10, or 30 mM. For monitoring the cell viability and functionality, the release of intracellular enzymes, parameters of glucose and nitrogen metabolism, as well as the liberation of inflammatory factors were measured daily. Upon termination of the bioreactor cultures, histological and immunofluorescence as well as mRNA analysis were performed to detect possible effects of APAP on the tissue integrity and gene expression of hepatic markers.

2. Materials and Methods

2.1. Bioreactor System

The microscale 3D bioreactor used in this study is based on a four-compartment hollow-fiber bioreactor technology described previously [15,17] and was further down-scaled resulting in a culture volume of about 100 μL for perfusion cell culture at microscale level. Figure 1 shows the bioreactor structure and the configuration of hollow-fiber capillaries in the device. The bioreactor housing (Figure 1A) is made of polyurethane and sized in credit card format. It disposes of tube connections for medium perfusion via two independent capillary systems (Medium I and Medium II), perfusion with an air/CO_2 mixture (Gas), and cell inoculation. A central cavity scaled at 1 cm in diameter harbors the capillary bed, which is made of four hollow-fiber layers. As shown in Figure 1B, each layer is composed of alternately arranged medium and gas capillaries, which serve for cell nutrition and oxygenation while providing an adhesion scaffold for the cells seeded in the extra-capillary space (cell compartment). The arrangement of capillary layers in a 45° angle to each other allows for counter-current medium perfusion of the capillary bed. Mass exchange between the capillary lumen and the cell compartment occurs via the pores of the used hydrophilic high-flux filtration membranes (3M, Neuss, Germany), while air/CO_2 exchange is mediated via hydrophobic membranes (Mitsubishi, Tokyo, Japan).

A Bioreactor structure

B Capillary configuration

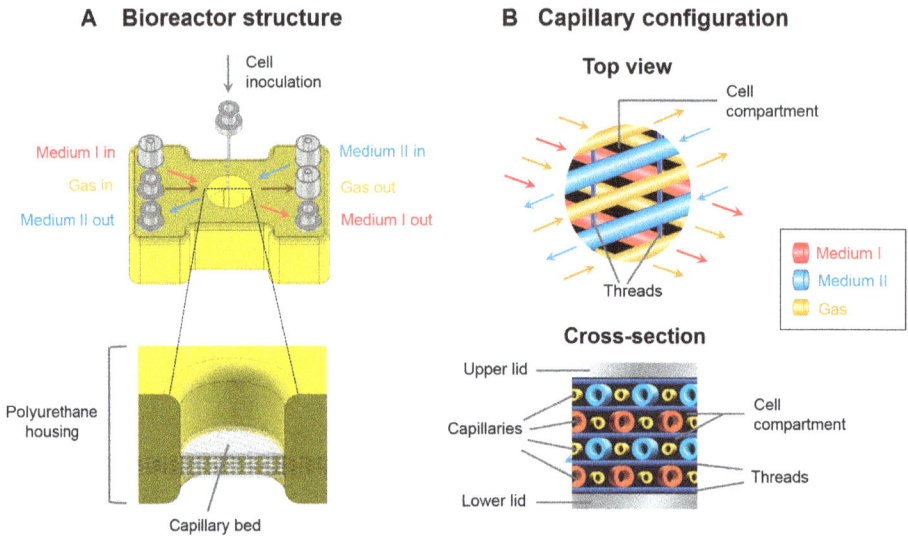

Figure 1. Schematic illustration of the structure of the microscale 3D liver bioreactor. (**A**) Outside view of the bioreactor with tube connections for medium in- and outflow via two independent capillary systems (Medium I and Medium II), perfusion with an air/CO_2 mixture (Gas), and a tube serving for cell inoculation into the extracapillary space (cell compartment); the figure below shows a section of the bioreactor housing with a central cavity containing the capillary bed; (**B**) Capillary structure of the bioreactor shown as top view (upper figure) and cross-section (lower figure). The capillary bed consists of four layers of hollow-fiber capillaries. Cells are seeded in the extra-capillary space (cell compartment). The layers form a 45° angle to each other to enable counter-current medium perfusion of the two capillary systems (red: Medium I, blue: Medium II). Synthetic threads (dark blue) integrated between each layer serve as spacers.

Synthetic threads made of polyethylenterephtalate are placed as spacers between the capillary layers. Thus, direct contacts between the capillaries, which could result in shunt formation and impaired mass exchange, are prevented.

Bioreactors are run in a perfusion circuit consisting of tubing for medium recirculation, medium feed, and medium outflow, as well as gas perfusion lines, as shown schematically in Figure 2. Medium flow rates are regulated by individual pumps for medium recirculation and medium feed, while electronically controlled gas valves (Vögtlin Instruments, Aesch, Switzerland) serve for regulation of air and CO_2 flow rates. The temperature in the bioreactor chamber is maintained at a constant level by means of software-controlled heating cartridges (HS Heizelemente GmbH, Fridingen, Germany).

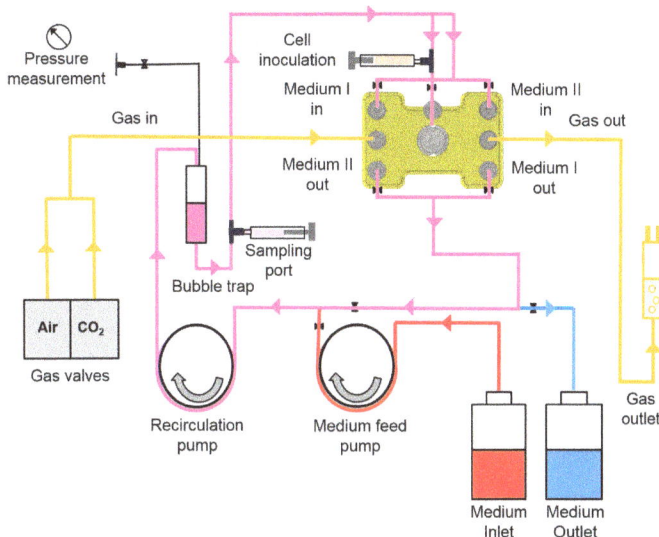

Figure 2. Schematic illustration of the bioreactor perfusion circuit. The bioreactor is integrated into a tubing circuit with continuously recirculating medium (purple). Fresh medium (red) is infused into the circuit via the medium feed pump while used medium (blue) is rinsed out from the circuit upon hydrostatic pressure increase. The tubing circuit contains a bubble trap connected with a line for pressure measurement and is equipped with connections for sample taking. Furthermore, the bioreactor is perfused with a defined air/CO_2 mixture (yellow), which is generated by means of electronically controlled gas valves.

2.2. Primary Human Liver Cell Isolation

Primary human liver cells were gained from tissue remaining from clinical partial liver resection. All patients gave their informed written consent before they participated in the study. The study was conducted in accordance with the Declaration of Helsinki, and the protocol was approved by the Ethics Committee of the Charité–Universitätsmedizin Berlin (EA2/026/09, dated 11 April 2013). Cell isolation was performed by collagenase digestion according to previously described protocols [18,19]. The obtained cell suspension was not further purified to avoid a loss of non-parenchymal cell fractions. The cell viability after isolation was 84.6 ± 7.0% as determined by their capacity to exclude trypan blue.

2.3. Bioreactor Operation

Primary human liver cells were seeded into each bioreactor at 10^7 cells per bioreactor. Liver cell bioreactors were perfused with Heparmed culture medium (Vito 143, Biochrom, Berlin, Germany), a modification of Williams' Medium E specifically developed for serum-free perfusion of high-density 3D liver cell cultures. The medium was supplemented prior to use with 20 IU/L insulin, 5 mg/L transferrin, 3 µg/L glucagon, 100,000 U/L penicillin, and 100 mg/L streptomycin (all purchased from Biochrom). Culture medium recirculated at a rate of 1 mL/min, while fresh medium was continuously fed into the perfusion circuit at a rate of 0.6 mL/h for the first 24 h followed by 0.2 mL/h until the end of culture. Used medium was removed at the same rate and flowed into the outlet vessel. The pH value in the recirculating medium was kept between 7.35 and 7.45 by adjusting the CO_2 concentration in the supplied gas mixture. Perfusion parameters and operation conditions for running the microscale bioreactors are listed in Table 1.

Table 1. Bioreactor perfusion parameters and operation conditions.

Parameter	Set Values during Operation
Recirculation rate	1 mL/min
Feed rate	0.6 mL/h (0–24 h)
	0.2 mL/h (from 24 h on)
Gas flow rate	4 mL/min
Concentration of CO_2 in supplied gas mixture	3–6% [1]
Temperature in bioreactor chamber	38 °C
pH value	7.35–7.45

[1] The CO_2 concentration was adjusted on demand to maintain a constant pH value in the system.

2.4. Clinical Chemistry Parameters

Parameters for assessment of the cell viability and functionality in the bioreactors were measured daily in samples from the culture perfusate. Activities of intracellular enzymes, including lactate dehydrogenase (LDH), aspartate aminotransferase (AST), alanine transaminase (ALT), and glutamate dehydrogenase (GLDH), as well as urea and ammonia concentrations were measured at Labor Berlin GmbH, Berlin, Germany, using automated clinical chemistry analyzers (Cobas® 8000, Roche Diagnostics GmbH, Mannheim, Germany). Glucose and lactate levels were analyzed by means of a blood gas analyzer (ABL 700, Radiometer, Copenhagen, Denmark). Concentrations of prostaglandin E2 (PGE2; Life Technologies, Carlsbad, CA, USA) and interleukin-6 (IL-6; Peprotech, Rocky Hill, NJ, USA) were determined using ELISA Kits according to the instructions of the manufacturers. The required sample volumes and variation coefficients are provided in Table 2.

Table 2. Sample volumes and variation coefficients of analyzed clinical chemistry parameters.

Parameter	Required Volume	Variation Coefficient
Glucose	100 μL in total	4%
Lactate		3%
LDH	250 μL in total	1%
AST		4%
ALT		2.9%
GLDH		0.8%
Urea		1%
Ammonia	250 μL	2.2%
PGE2	200 μL	n.a.
IL-6	200 μL	n.a.

2.5. Acetaminophen (APAP) Application

On the third day of culture, APAP (Sigma-Aldrich, St.-Louis, MO, USA) was added at a final concentration of 5, 10, or 30 mM. The drug was dissolved in methanol, followed by methanol evaporation and dissolution of the substance in culture medium. APAP incubation was initiated by adding 1 mL of a 7× concentrated solution of the drug into the perfusion circuit via the bubble trap (bolus application), to reach the desired final solution of the drug in the recirculation circuit. Subsequently, fresh medium containing APAP at the respective concentration of 5, 10, or 30 mM was continuously infused into the perfusion circuit. The control bioreactor was treated equally, but without adding the drug.

2.6. Histological and Immunofluorescence Analysis

Upon termination of bioreactor cultures on day 6 of culture, bioreactors were opened and sections of the capillary bed containing the cell material were taken. Histological slides were prepared from fixed samples and subjected to hematoxylin-eosin (HE) and immunofluorescence staining as described previously [15]. Double-staining of antigens was performed using monoclonal mouse anti-cytochrome P450 (CYP) 1A2 antibodies (AB) provided by Santa Cruz (Santa Cruz, CA, USA), combined with monoclonal rabbit anti-cytokeratin 18 (CK18) AB (Abcam, Cambridge, UK); monoclonal mouse anti-CK18 AB (Santa Cruz), combined with polyclonal rabbit anti-vimentin AB (Santa Cruz); or monoclonal mouse anti-multidrug resistance protein 2 (MRP2) AB (Abcam) in combination with polyclonal rabbit anti-CYP3A4 AB (Abcam). As secondary AB, fluorochrome-coupled goat anti-mouse IgG 488 AB (Life Technologies) and goat anti-rabbit IgG 594 AB (Life Technologies) were used. Counterstaining of nuclei was performed using bisbenzimide H 33342 trihydrochloride (Hoechst 33342, Sigma-Aldrich).

2.7. qRT-PCR

Total RNA was obtained from human liver cells gained from the bioreactors after termination of the cultures on day 6. The RNA was extracted using the TRIzol® Reagent (Life Technologies) according to the manufacturer's instructions. Afterwards, genomic DNA was digested using the RNase-free DNase-Set (Qiagen, Hilden, Germany). Subsequent cDNA synthesis and quantitative real-time PCR (qRT-PCR) were performed as described elsewhere [20] using human-specific primers and probes (TaqMan Gene Expression Assay system, Life Technologies, Table 3). The expression of specific genes was normalized to that of the housekeeping gene glyceraldehyde-3-phosphate dehydrogenase (*GAPDH*) and fold changes of expression levels were calculated with the $\Delta\Delta C_t$ method [21].

Table 3. Applied Biosystems TaqMan Gene Expression Assays®.

Gene Symbol	Gene Name	Assay ID
AIFM1	Apoptosis-inducing factor, mitochondria-associated, 1	Hs00377585_m1
CASP3	Caspase 3, apoptosis-related cysteine peptidase	Hs00234387_m1
CPS1	Carbamoyl-phosphate synthase 1, mitochondrial	Hs00157048_m1
CYP1A2	Cytochrome P450, family 1, subfamily A, polypeptide 2	Hs00167927_m1
CYP2E1	Cytochrome P450, family 2, subfamily E, polypeptide 1	Hs00559368_m1
GAPDH	Glyceraldehyde-3-phosphate dehydrogenase	Hs03929097_g1
GSTO2	Glutathione S-transferase omega 2	Hs01598184_m1

2.8. Statistics

Four independent experiments were performed with cells from different donors ($N = 4$, unless stated otherwise). Statistical analyses were performed using GraphPad Prism 5.0 for Windows (GraphPad Software, San Diego, CA, USA). Results are provided as mean ± standard error of the mean (SEM). The influence of the drug dose (day 3–day 6) on clinical chemistry parameters in comparison to the control was analyzed by calculating the area under curve (AUC) of values during the drug application interval. The AUCs between day 3 and day 6 of the groups treated with different APAP concentrations were compared with those of untreated control cultures by means of one-way ANOVA with Dunnett's multiple comparison test. The same test was used for statistical evaluation of gene expression data. The group treated with 30 mM APAP was not included in the statistical analysis of gene expression data, since RNA in sufficient quality and quantity was only gained from one culture in this group.

3. Results

3.1. Clinical Chemistry Parameters

Clinical chemistry parameters revealed a dose-dependent effect of APAP on metabolic functions of primary human liver cells maintained in perfused microscale bioreactors (Figure 3).

The time-course of glucose production (Figure 3A) showed stable values with some fluctuations in control bioreactors or those treated with 5 mM APAP, while a clear decrease upon drug application from day 3 onwards was observed in bioreactors exposed to 10 or 30 mM APAP. Lactate production rates (Figure 3B) showed a steadily increasing course in the control group, while bioreactors exposed to 5 or 10 mM APAP remained on a constant level, and the group exposed to 30 mM APAP was characterized by a sharp decline. The comparison of AUCs of lactate values following drug application revealed a significant difference between the 30 mM APAP-treated group and the control group ($p < 0.01$).

The time-course of ammonia release was determined as an indicator for the cells' capacity of nitrogen elimination (Figure 3C). After an initial peak on the first culture day, control bioreactors and those treated with 5 mM APAP showed stable values on a basal level. In contrast, a distinct increase was observed in bioreactors upon exposure to 10 or 30 mM APAP, with significantly ($p < 0.0001$) increased AUCs as compared with the control group. Urea production rates showed a mild, but not significant decrease at 30 mM APAP, while lower drug concentrations did not affect urea levels as compared to untreated control bioreactors (Figure 3D).

Release rates of the intracellular enzymes LDH and AST, indicating disturbed cell integrity and membrane leakage, showed a similar time-course in all experimental groups, characterized by a peak immediately after cell inoculation, which was followed by basal levels from day 3 onwards (Figure 3E,F). The enzymes ALT and GLDH showed a similar time course (data available at http://doi.org/10.5281/zenodo.1169306 (clinical chemistry parameters)). The administration of APAP had no effect on enzyme release rates.

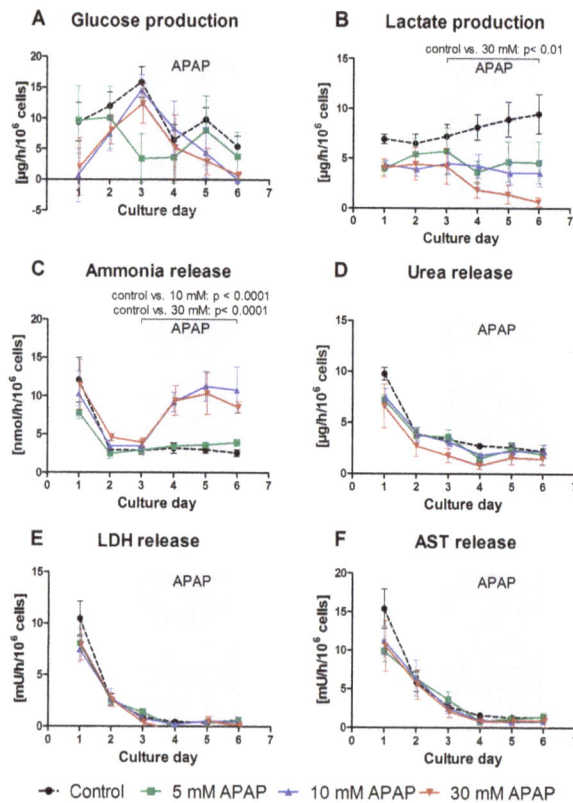

Figure 3. Time-courses of clinical parameters in bioreactors treated with 5 mM, 10 mM or 30 mM acetaminophen (APAP) in comparison to untreated bioreactors used as control group. The figure shows

the course of glucose (**A**) and lactate (**B**) production, ammonia (**C**) and urea (**D**) release, as well as liberation of lactate dehydrogenase (LDH, (**E**)) and aspartate aminotransferase (AST, (**F**)). APAP was continuously introduced from day 3 throughout day 6 of culture. Values were normalized to 10^6 inoculated cells. Data are shown as means $\pm$ SEM ($n = 4$; control $n = 6$). The influence of the drug dose (day 3–day 6) on the metabolic activity of the cells in comparison to the control was analyzed by means of one-way ANOVA with Dunnett's multiple comparison test, using the AUCs from day 3 until day 6. Significant changes are indicated in the graphs. Underlying data are available at http://doi.org/10.5281/zenodo.1169306 (Clinical_chemistry_parameters).

The release of inflammatory factors was selectively affected by different APAP concentrations (Figure 4).

Release rates of the prostaglandin PGE2 (Figure 4A) showed a peak immediately after cell isolation followed by a rapid decline to basal levels. Upon APAP exposure, PGE2 release rates were characterized by a significant ($p < 0.05$) increase in bioreactors treated with 30 mM APAP as compared with the control group, indicating induction of PGE2 secretion at high APAP doses. In contrast, the cytokine IL-6 showed a general decrease in values in all groups during the culture course, and AUCs upon APAP incubation were similar in APAP-treated and untreated cultures (Figure 4B).

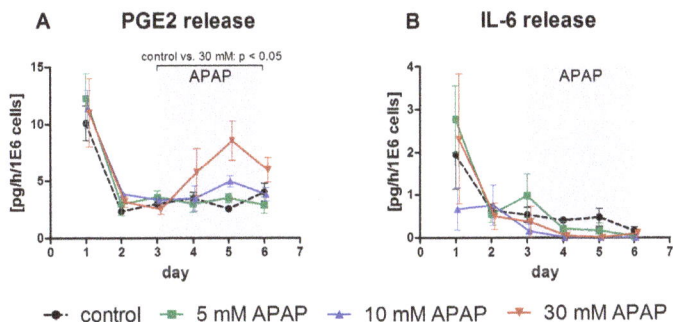

Figure 4. Time-courses of inflammatory factors in bioreactors treated with 5 mM, 10 mM, or 30 mM acetaminophen (APAP) in comparison to untreated bioreactors used as control group. The figure shows release rates of prostaglandin E2 (PGE2, (**A**)) and interleukin 6 (IL-6, (**B**)). APAP was continuously introduced from day 3 throughout day 6 of culture. Values were normalized to 10^6 inoculated cells. Data are shown as means $\pm$ SEM ($n = 3$; control $n = 5$). The influence of the drug dose (day 3–day 6) on the metabolic activity of the cells in comparison to the control was analyzed by means of one-way ANOVA with Dunnett's multiple comparison test, using the AUCs from day 3 until day 6. Significant changes are indicated in the graphs. Underlying data are available at http://doi.org/10.5281/zenodo.1169306 (Clinical_chemistry_parameters).

3.2. Gene Expression Analysis

The expression of genes involved in drug metabolism, antioxidant reactions, urea synthesis, and apoptosis was influenced individually by APAP exposure (Figure 5).

The genes encoding for *CYP1A2* (Figure 5A) and *CYP2E1* (Figure 5B) were strongly reduced in cultures exposed to 5 mM APAP as compared to the untreated control bioreactors, followed by a stepwise increase at higher APAP concentrations. While *CYP1A2* expression was lower in all APAP-treated groups than in the control, the expression of *CYP2E1* increased by 20-fold in the group treated with 30 mM APAP. The genes encoding for carbamoyl phosphate synthetase I (*CPS1*, Figure 5C) and glutathione S-transferase omega 2 (*GSTO2*, Figure 5D) showed a similar expression pattern, characterized by a 25-fold (*CPS1*, $p < 0.05$) resp. 4.5-fold (*GSTO2*) reduction in bioreactors treated with 5 mM APAP, and a successive increase at higher drug concentrations, without reaching the values of the control group. A different effect of APAP on gene expression was observed for the genes associated with apoptosis, namely caspase 3, apoptosis-related cysteine peptidase (*CASP3*, Figure 5E), and apoptosis-inducing factor, mitochondria-associated, 1 (*AIFM1*, Figure 5F). *CASP3* showed a slight increase in expression at 5 mM APAP, which was followed by a successive decline in the groups exposed to 10 or 30 mM APAP, whereas *AIFM1* expression was reduced in all APAP-treated groups, with significantly lower values in bioreactors treated with 5 or 10 mM APAP ($p < 0.0001$).

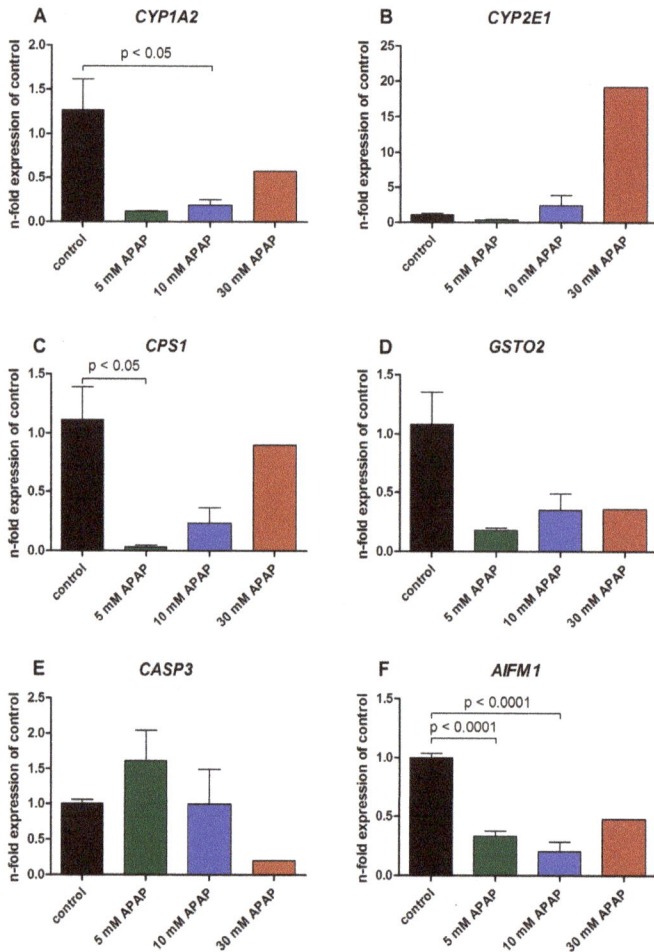

Figure 5. Gene expression analysis of primary human liver cells after culture in bioreactors treated with 5 mM, 10 mM or 30 mM acetaminophen (APAP) in comparison to untreated bioreactors used as control group. The figure shows the gene expression of cytochrome P450 family 1, subfamily A, polypeptide 2 (*CYP1A2*, (**A**)), cytochrome P450 family 2, subfamily E, polypeptide 1 (*CYP2E1*, (**B**)), carbamoyl phosphate synthetase I (*CPS1*, (**C**)), glutathione S-transferase omega 2 (*GSTO2*, (**D**)), caspase 3, apoptosis-related cysteine peptidase (*CASP3*, (**E**)) and apoptosis-inducing factor, mitochondria-associated, 1 (*AIFM1*, (**F**)). Expression data were normalized to the house-keeping gene glyceraldehyde-3-phosphate dehydrogenase and were calculated relative to the untreated control using the ΔΔCt-method. Data are shown as means ± SEM (control *n* = 4; 5 mM APAP *n* = 2; 10 mM APAP

n = 3, 30 mM APAP n = 1). Differences between the control and 5 mM or 10 mM APAP were calculated using one-way ANOVA with Dunnett's multiple comparison test and significant changes are indicated in the graphs. Underlying data are available at http://doi.org/10.5281/zenodo.1169306 (qRT_PCR).

3.3. Histological and Immunohistochemical Analysis

Histological investigation and immunofluorescence labeling of hepatic antigens in untreated cultures (control) or those exposed to 10 or 30 mM APAP showed a clear effect of APAP on the tissue organization and distribution pattern of cell-specific markers in bioreactor cultures (Figure 6).

As shown by HE staining (Figure 6A–C), primary human liver cells cultured in control bioreactors were associated in tissue-like cell aggregates between the hollow-fiber capillaries. Cell clusters contained primarily hepatocytes characterized by a large cytoplasm and a round or polygonal shape. The majority of cells appeared morphologically intact. Upon treatment with 10 or 30 mM APAP, cell aggregates and cell–cell connections were partly dissolved, resulting in the occurrence of numerous isolated cells. Most cells displayed a small and condensed cytoplasm and a lack of demarcation of cell nuclei, indicating necrotic and/or apoptotic processes.

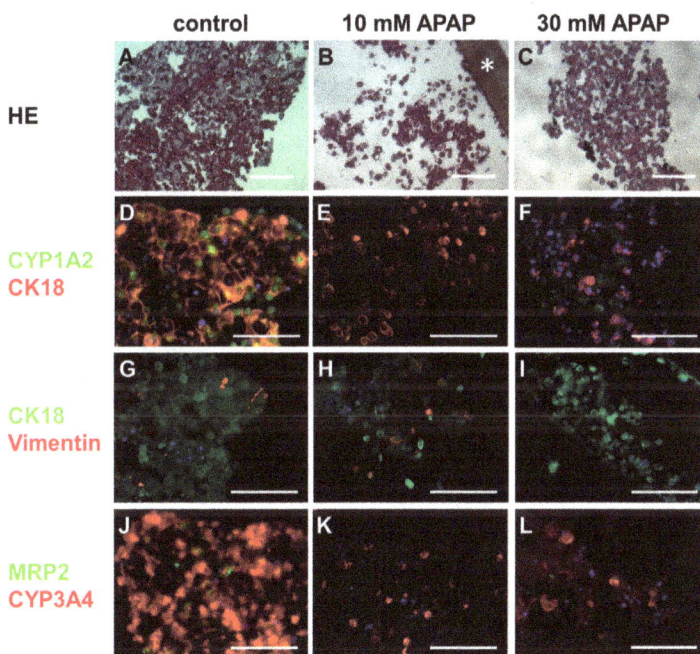

Figure 6. Histological and immunohistochemical staining of primary human liver cells after culture in untreated bioreactors (control) or those treated with 10 mM or 30 mM acetaminophen (APAP). The figure shows the hematoxylin-eosin stain (HE, (**A–C**)); double staining of cytochrome P450 family 1, subfamily A, polypeptide 2 (CYP1A2) and cytokeratin 18 (CK18) (**D–F**), CK18, and vimentin (**G–I**); and double staining of multidrug resistance-associated protein 2 (MRP2) and cytochrome P450 family 3, subfamily A, polypeptide 4 (CYP3A4) (**J–L**). The asterisk in (**B**) marks a hollow-fiber capillary membrane. Nuclei were counter-stained with Hoechst 33342 (blue). Scale bars correspond to 100 μm. Source pictures are available at http://doi.org/10.5281/zenodo.1169306 (Histology_Immnunofluorescence).

Immunofluorescence staining confirmed the finding of cell damage and partial disintegration of cell aggregates upon APAP application. In control cultures, the hepatocyte-specific markers CK18 and CYP1A2 (Figure 6D–F) mostly showed an evenly distributed staining. The cytoskeletal marker CK18 was primarily expressed at cell margins, forming a network-like staining pattern, while CYP1A2 was mainly localized in the cytoplasm. In cultures treated with 10 or 30 mM APAP sparse and irregular staining of CK18 and CYP1A2 was observed. Staining of nuclei with Hoechst 33,342 revealed a condensed cytoplasm of most cells and furthermore, the occurrence of isolated nuclei devoid of cytoplasm, as an additional indication of cell death. Double-staining of CK18 and vimentin (Figure 6G–I) showed the presence of some non-parenchymal cells (vimentin-positive) between hepatocytes (CK18-positive), both in control bioreactors and those treated with 10 mM APAP. In contrast, no vimentin-positive cells were detected in cultures exposed to 30 mM APAP, indicating a loss of non-parenchymal cells. Expression of CYP3A4 (Figure 6J–L) was observed in the cytoplasm of most cells in control bioreactors and was still expressed in part of the cells after APAP treatment. The biliary transporter MRP2 being localized in plasma membranes of adjacent cells was detected in around 40–50% of the cells in control cultures. In bioreactors subjected to 10 or 30 mM APAP, the fraction of MRP2 positive cells was decreased to less than 10%.

4. Discussion

In order to precisely predict the hepatotoxicity of compounds during drug development, validated human in vitro models are necessary. Models based on a 3D environment have proven to more accurately reflect the human body compared to conventional 2D models [6]. Various microfluidic culture systems were developed to minimize the amounts of cells and culture materials [12,13]. The microscale 3D bioreactor used in this study is based on an existing four-compartment hollow-fiber technology [14–17] and was down-scaled to a cell compartment volume of 100 µL and a cultivated cell number of 10 million primary human liver cells.

To demonstrate the suitability of the microscale bioreactor system for hepatotoxicity studies, APAP was applied as a gold-standard test substance in concentrations of 5, 10, or 30 mM over a time period of three days. APAP toxicity manifests in many different ways as reviewed by Hinson et al. (2010), such as glutathione depletion, formation of toxic protein adducts, enzyme and cytokine release, and histological alterations [22]. In this study, the focus was on clinical chemistry parameters allowing regular evaluation of APAP toxicity during culture, followed by end-point analyses allowing the judgement of alterations in tissue integrity as well as protein and gene expression.

The time course of clinical chemistry parameters measured in the culture perfusate was generally characterized by an initial peak on the first day of culture. This increase can be ascribed to the cell isolation process, which leads to high levels of cell stress causing the release of intracellular enzymes and metabolites. In addition, disruption of cell–cell and cell–matrix contacts and consequently the loss of cell polarization were described [23]. Liver cells may also become pre-activated due to reperfusion injury associated with oxidative stress [24]. Both disruption of tissue integrity and activation of inflammatory signaling, are associated with dedifferentiation processes starting already during isolation [5].

In concordance with other studies [7,25] we observed dose-dependent toxic effects of APAP on primary human liver cells cultured in the device. Glucose and lactate production rates measured as parameters for energy metabolism showed a dose-dependent decrease from day 3 (beginning of APAP dosing) onwards indicating an impaired cell viability and functionality. The suitability of glucose consumption and lactate production to detect drug-induced changes in cell viability of hepatic cell cultures was previously shown [26]. In contrast to assays based on substrate conversion, for example, MTT or cell titer blue assay, measurement of glucose and lactate allows a regular monitoring of cell activity over time without intervening into the cell metabolism.

Additionally, the enzyme release was measured as an indicator for necrosis or for secondary necrosis following apoptosis [27,28]. Serum levels of the cytoplasmic enzyme LDH are routinely measured in clinical settings in order to assess cell damage within pathological processes, including liver injury [29,30]. AST is distributed both in the cytoplasm and mitochondria of hepatocytes [31]; mild cell injury causes the release of cytosolic enzymes, whereas severe liver damage leads to the release of both, cytoplasmic and mitochondrial enzymes. All experimental groups showed an initial peak of enzyme values on day 1 in consistence with other studies [32]. The absence of an increase in enzyme release during the APAP treatment period is in accordance with previous results investigating diclofenac toxicity in 3D bioreactors [15]. Similarly, periodic treatment of hepatocyte cultures with 18.6 mM APAP for 20 days had no significant effect on LDH release [33]. The absence of response in hepatic enzyme release may be explained by the exhaustion of cytosolic enzyme stores in the initial culture phase, which can be ascribed to cell stress during cell isolation. In a study investigating liver enzyme concentrations in the cytosol and in the supernatant of primary human hepatocytes exposed to APAP, the amount of secreted enzyme relative to the total enzyme content showed a significant increase upon drug application [25]. Hence, the determination of both extracellular and intracellular LDH activity could provide more conclusive results on the actual effect of APAP on enzyme release. However, this would require daily lysis of cell samples, which is difficult to realize in complex 3D culture systems.

Ammonia, a product of amino acid metabolism, is toxic in high concentrations and is therefore converted to urea by hepatocytes. Hence, through the analysis of urea synthesis and ammonia depletion, conclusions can be drawn regarding the functionality of cultured hepatocytes [34]. The observed initial increase in urea and ammonia production rates can be related to cell stress due to cell isolation, in accordance with the observed peaks in enzyme release. The finding of significantly increased ammonia release rates after dosing with 10 or 30 mM APAP from day 3 onwards indicate a dose-dependent influence of APAP on the nitrogen metabolism of the cells. This finding is supported by the fact that APAP toxicity causes mitochondrial dysfunction [25], since part of the enzymes involved in the urea cycle are located in the mitochondria. Moreover, these observations emphasize the suitability of ammonia as a sensitive parameter for hepatocyte functionality and hepatic drug toxicity.

The inflammatory factors PGE2 and IL-6 showed a different course upon drug application. A dose-dependent increase from day 3 (beginning of APAP dosing) onwards was detected for PGE2, a pro-inflammatory factor with immunosuppressive activity, indicating an accumulation of APAP-induced inflammation. Since PGE2 is typically produced by cell types such as endothelial cells and cells of the immune system [35], the finding of an increased PGE2 release confirms the presence of non-parenchymal cells in the liver bioreactor. In contrast, the pro-inflammatory cytokine IL-6, which is secreted by different liver cell types, including hepatocytes [36], showed no significant response to APAP exposure. PGE2 inhibits the production of IL-6 as a mechanism of limiting excessive immune reactions [37], which may be the reason for the lack of IL-6 response to APAP treatment.

Since production rates of metabolic parameters were normalized to the initial cell number, the results from the medium analysis represent the combined effect of cell number and functionality. Thus, values do not allow a distinction between a change either of cell number or of cell activity in the applied culture system.

In the liver, APAP is metabolized mainly by *CYP2E1* to the toxic N-acetyl-*p*-benzochinonimin (NAPQI) [38,39]. Gene expression analysis of the cells subsequent to APAP exposure in comparison to untreated control bioreactors revealed a dose-dependent increase in *CYP2E1* expression especially for 30 mM APAP. This result indicates that high concentrations of APAP lead to an immediate upregulation of *CYP2E1*, which can be seen as a mechanism to accelerate APAP metabolism. In addition, we observed a decrease in gene expression of *CYP1A2* upon APAP treatment compared to the control bioreactor, though less pronounced at increasing APAP concentrations. The contribution of CYP1A2 to APAP metabolism is controversially discussed. While studies considering rat and human liver microsomes [40,41] or recombinant human CYP P450 enzymes [42] showed that CYP1A2 is involved in the metabolism of APAP, other publications, performed in adult human volunteers, reported no direct association of CYP1A2 with APAP depletion [38,43]. These contradictory findings may be caused by different conditions in in vitro experiments as compared with the in vivo situation.

The glutathione S-transferases (GSTs) are enzymes catalyzing the neutralization of free radicals and active drug components using glutathione as reducing agent, which is the main step in phase 2 detoxification [44]. In APAP metabolism, GSTs catalyze the formation of a NAPQI-glutathione adduct, which is then primarily secreted into bile by passing the apical membrane transporter protein MRP2 [45]. Treatment of hepatocytes with APAP resulted in a decrease of *GSTO2* gene expression as compared to the control. Similar findings were reported by Wang et al. (2017), who observed a reduction of GST activities in a mouse model of APAP-induced liver injury [46]. The observed decrease in gene expression of *CPS-1*, an enzyme involved in the production of urea, is in accordance to our observations of increased ammonia release rates in the bioreactors treated with APAP, and further supports the assumption that APAP effects the nitrogen metabolism.

In contrast, the apoptosis-associated gene *CASP3* showed an increase in expression for the bioreactor treated with 5 mM APAP in comparison to the untreated controls, while no change was observed at 10 mM APAP and a decrease was detected after exposure to 30 mM APAP. This observation might be explained by a shift to necrotic cell death, in accordance with findings by Au and colleagues, who observed a transition from apoptosis to necrosis between 10 and 20 mM APAP exposure using HepG2 cells [47]. However, another apoptosis-associated factor, *AIFM1*, revealed reduced expression values for all drug-treated bioreactors as compared to the control. The pro-apoptotic function of AIFM1 is based on the activation of a caspase-independent pathway upon apoptotic stimuli, whereas its anti-apoptotic function is part of the regular mitochondria metabolism via NADH oxidoreduction [48]. As APAP treatment results in mitochondrial dysfunction, it might consequently also lead to a reduced gene expression of *AIFM1*. Since APAP is known to induce both necrosis and apoptosis [49], a more detailed characterization of the type of cell death would be of interest in future studies to differentiate between necrosis and apoptosis.

Histological and immunohistochemical analyses revealed that the cells cultured in control bioreactors were reorganized in tissue-like formations, which resembled those observed in larger scale bioreactors with higher initial cell amounts [14,16]. Typical structural and functional markers of hepatocytes, including CK18, CYP1A1 and CYP3A4 were regularly detected and showed an in vivo-like distribution pattern. Staining of MRP2, which is the main transporter for biliary excretion of acetaminophen sulfate [50,51], indicates cell polarization with formation of bile canaliculi. The characterization of the cells by means of cell-specific markers showed that in addition to hepatocytes, the aggregates also comprised non-parenchymal cells identified by vimentin staining. Since the non-parenchymal cells of the liver play a major role in drug-induced liver injury [52], the presence of these cells in liver models is critical to assess complex drug effects mediated by different liver cell populations.

Exposure to APAP at a concentration of 10 or 30 mM resulted in partial disintegration of cell aggregates and loss of cell integrity. In particular, MRP2 was rarely detectable in APAP-treated bioreactors indicating a depolarization of hepatocytes. This is in line with findings by Bhise and colleagues, who showed a massive reduction of MRP2 immunostaining after treatment of hepatic spheroids composed of hepatoma cells with 15 mM APAP [53]. The reduction of membrane transporters may lead to impaired excretion of APAP metabolites and therefore cause accumulation of toxic products in the cells, if glutathione is not sufficiently available.

In summary, we were able to detect dose-dependent hepatotoxic effects of 5 to 30 mM APAP on primary human hepatocytes cultured in the microscale 3D bioreactor. Au and colleagues identified hepatotoxic influences of APAP at 10 mM using spheroids comprising HepG2 cells and fibroblasts cultured on a microfluidic platform [47]. Other microfluidic studies using HepG2 cells found a reduction of cell viability by more than 50% upon treatment with 15 mM APAP [53] resp. 20 mM APAP [54]. However, in the human body, plasma concentrations of 0.5 to 3 mM APAP were observed in overdose scenarios [55]. A potential reason for the discrepancy between

Bioengineering **2018**, *5*, 24

toxic APAP concentrations in vivo and in vitro can be seen in the contribution of systemic influences, such as nutritional status [56] and blood cells [57], to APAP toxicity. In addition, the non-parenchymal liver cells, such as sinusoidal endothelial cells [58] and Kupffer cells [59], have been shown to play a role in APAP toxicity. Although in the present study the obtained mixture of primary liver cells after enzymatic digestion of the organ was used without further purification of hepatocytes, the number of non-parenchymal cells might not have been sufficient to correctly imitate the human in vivo liver.

Hence, future studies would be of interest to further investigate the role of individual liver cell populations in APAP metabolism and toxicity. The addition of non-parenchymal liver cells to the microscale 3D liver culture in ratios comparable to the in vivo situation could increase its sensitivity for APAP toxicity. Methods for isolation of endothelial cells, Kupffer cells and stellate cells from human liver tissue were recently described [18] and can be used to provide defined amounts of these cells for human liver cell models. Other microfluidic systems attempt to recapitulate the microarchitecture of the liver sinusoid by providing several compartments for the different liver cell types [60,61]. A further important factor in drug susceptibility might be the oxygen concentration, as indicated by results from co-cultures of primary rat hepatocytes and fibroblasts, which proved to be more sensitive to APAP exposure in low-oxygen regions [62]. Thus, the creation of physiological oxygen gradients might increase the predictive power of in vitro drug effects. To assess systemic effects mediated by other organs, integration of the microscale 3D liver bioreactor into a multi-organ platform would be an attractive approach. In this context, the structure of the bioreactor provides suitable conditions for realization of microscale systems integrating various organ constructs, such as kidney, heart, and lung.

A general limitation of microscale systems can be seen in the small amount of cell material available for end-point analyses such as immunohistochemistry, qPCR, or Western blots. Thus, studies investigating microscale liver tissues often show results from only one end-point analysis, mostly immunohistochemistry or Western blotting [32,47,53,54,62]. Hence, the implementation of analytic methods allowing analyses from minimal cell numbers is required to generate a larger range of data from microscale systems.

5. Conclusions

In conclusion, the results from APAP application in this study demonstrate that the microscale 3D liver bioreactor provides a useful tool to detect hepatotoxic effects of drugs in a perfused human in vitro culture environment. Our observations emphasize the potential of clinical chemistry parameters, such as lactate production and ammonia release, as sensitive parameters for monitoring dose-dependent hepatotoxic effects throughout the culture period. The analysis of inflammatory factors showed that mainly PGE2 was affected by APAP exposure in the model. The toxic effect of APAP in the in vitro model was confirmed by end-point analyses, including histological and immunohistochemical evaluation and PCR analysis. In order to increase the sensitivity of the present microscale 3D liver bioreactor for toxicity studies at physiologically relevant drug concentrations, co-cultures supplemented with different non-parenchymal cell types at physiological ratios, and also creation of defined oxygen gradients could be applied in future studies.

Supplementary Materials: Raw data underlying the presented figures are available online on the data repository "Zenodo" under http://doi.org/10.5281/zenodo.1169306 with the doi:10.5281/zenodo.1169306.

Acknowledgments: The work for the study was supported, in part, by the Defense Threat Reduction Agency (program: Integration of Novel Technologies for Organ Development and Rapid Assessment of Medical Countermeasures, Interagency Agreement DTRA100271A5196), USA, and in part by the German Ministry for Education and Research (BMBF) within the funding network "Virtual Liver" (FKZ 0315741).

Author Contributions: Nora Freyer and Selina Greuel analyzed the data and wrote the manuscript; Fanny Knöspel conceived and designed the experiments and evaluated the data; Florian Gerstmann and Lisa Storch performed the experiments and contributed to data analysis; Georg Damm and Daniel Seehofer were responsible for isolation, processing, and quality analysis of fresh primary human liver cells; Jennifer Foster Harris and Rashi Iyer contributed to the conception and design of experiments and writing of the manuscript; Frank Schubert was responsible for technical developments and design of bioreactor prototypes; and Katrin Zeilinger contributed to the design of experiments, evaluation of results and writing of the manuscript.

Conflicts of Interest: The authors declare no conflict of interest.

References

1. Bhakuni, G.S.; Bedi, O.; Bariwal, J.; Deshmukh, R.; Kumar, P. Animal models of hepatotoxicity. *Inflamm. Res.* **2016**, *65*, 13–24. [CrossRef] [PubMed]
2. Maes, M.; Vinken, M.; Jaeschke, H. Experimental models of hepatotoxicity related to acute liver failure. *Toxicol. Appl. Pharmacol.* **2016**, *290*, 86–97. [CrossRef] [PubMed]
3. Sharma, V.; McNeill, J.H. To scale or not to scale: The principles of dose extrapolation. *Br. J. Pharmacol.* **2009**, *157*, 907–921. [CrossRef] [PubMed]
4. Olson, H.; Betton, G.; Robinson, D.; Thomas, K.; Monro, A.; Kolaja, G.; Lilly, P.; Sanders, J.; Sipes, G.; Bracken, W.; et al. Concordance of the toxicity of pharmaceuticals in humans and in animals. *Regul. Toxicol. Pharmacol.* **2000**, *32*, 56–67. [CrossRef] [PubMed]
5. Elaut, G.; Henkens, T.; Papeleu, P.; Snykers, S.; Vinken, M.; Vanhaecke, T.; Rogiers, V. Molecular mechanisms underlying the dedifferentiation process of isolated hepatocytes and their cultures. *Curr. Drug Metab.* **2006**, *7*, 629–660. [CrossRef] [PubMed]
6. Elliott, N.T.; Yuan, F. A review of three-dimensional in vitro tissue models for drug discovery and transport studies. *J. Pharm. Sci.* **2011**, *100*, 59–74. [CrossRef] [PubMed]
7. Schyschka, L.; Sánchez, J.J.; Wang, Z.; Burkhardt, B.; Müller-Vieira, U.; Zeilinger, K.; Bachmann, A.; Nadalin, S.; Damm, G.; Nussler, A.K. Hepatic 3D cultures but not 2D cultures preserve specific transporter activity for acetaminophen-induced hepatotoxicity. *Arch. Toxicol.* **2013**, *87*, 1581–1593. [CrossRef] [PubMed]
8. Bell, C.C.; Hendriks, D.F.; Moro, S.M.; Ellis, E.; Walsh, J.; Renblom, A.; Fredriksson Puigvert, L.; Dankers, A.C.; Jacobs, F.; Snoeys, J.; et al. Characterization of primary human hepatocyte spheroids as a model system for drug-induced liver injury, liver function and disease. *Sci. Rep.* **2016**, *6*, 25187. [CrossRef] [PubMed]
9. Ohkura, T.; Ohta, K.; Nagao, T.; Kusumoto, K.; Koeda, A.; Ueda, T.; Jomura, T.; Ikeya, T.; Ozeki, E.; Wada, K.; et al. Evaluation of human hepatocytes cultured by three-dimensional spheroid systems for drug metabolism. *Drug Metab. Pharmacokinet.* **2014**, *29*, 373–378. [CrossRef] [PubMed]
10. Dash, A.; Simmers, M.B.; Deering, T.G.; Berry, D.J.; Feaver, R.E.; Hastings, N.E.; Pruett, T.L.; LeCluyse, E.L.; Blackman, B.R.; Wamhoff, B.R. Hemodynamic flow improves rat hepatocyte morphology, function, and metabolic activity in vitro. *Am. J. Physiol. Cell Physiol.* **2013**, *304*, C1053–C1063. [CrossRef] [PubMed]
11. De Bartolo, L.; Salerno, S.; Curcio, E.; Piscioneri, A.; Rende, M.; Morelli, S.; Tasselli, F.; Bader, A.; Drioli, E. Human hepatocyte functions in a crossed hollow fiber membrane bioreactor. *Biomaterials* **2009**, *30*, 2531–2543. [CrossRef] [PubMed]
12. Aziz, A.U.R.; Geng, C.; Fu, M.; Yu, X.; Qin, K.; Liu, B. The role of microfluidics for organ on chip simulations. *Bioengineering* **2017**, *4*, 39. [CrossRef] [PubMed]
13. Gupta, N.; Liu, J.R.; Patel, B.; Solomon, D.E.; Vaidya, B.; Gupta, V. Microfluidics-based 3D cell culture models: Utility in novel drug discovery and delivery research. *Bioeng. Transl. Med.* **2016**, *1*, 63–81. [CrossRef] [PubMed]
14. Zeilinger, K.; Schreiter, T.; Darnell, M.; Söderdahl, T.; Lübberstedt, M.; Dillner, B.; Knobeloch, D.; Nüssler, A.K.; Gerlach, J.C.; Andersson, T.B. Scaling down of a clinical three-dimensional perfusion multicompartment hollow fiber liver bioreactor developed for extracorporeal liver support to an analytical scale device useful for hepatic pharmacological in vitro studies. *Tissue Eng. Part C* **2011**, *17*, 549–556. [CrossRef] [PubMed]
15. Knöspel, F.; Jacobs, F.; Freyer, N.; Damm, G.; De Bondt, A.; van den Wyngaert, I.; Snoeys, J.; Monshouwer, M.; Richter, M.; Strahl, N.; et al. In vitro model for hepatotoxicity studies based on primary human hepatocyte cultivation in a perfused 3D bioreactor system. *Int. J. Mol. Sci.* **2016**, *17*, 584. [CrossRef] [PubMed]
16. Lübberstedt, M.; Müller-Vieira, U.; Biemel, K.M.; Darnell, M.; Hoffmann, S.A.; Knöspel, F.; Wönne, E.C.; Knobeloch, D.; Nüssler, A.K.; Gerlach, J.C.; et al. Serum-free culture of primary human hepatocytes in a miniaturized hollow-fibre membrane bioreactor for pharmacological in vitro studies. *J. Tissue Eng. Regen. Med.* **2015**, *9*, 1017–1026. [CrossRef] [PubMed]
17. Hoffmann, S.A.; Müller-Vieira, U.; Biemel, K.; Knobeloch, D.; Heydel, S.; Lübberstedt, M.; Nüssler, A.K.; Andersson, T.B.; Gerlach, J.C.; Zeilinger, K. Analysis of drug metabolism activities in a miniaturized liver cell bioreactor for use in pharmacological studies. *Biotechnol. Bioeng.* **2012**, *109*, 3172–3181. [CrossRef] [PubMed]

18. Kegel, V.; Deharde, D.; Pfeiffer, E.; Zeilinger, K.; Seehofer, D.; Damm, G. Protocol for isolation of primary human hepatocytes and corresponding major populations of non-parenchymal liver cells. *J. Vis. Exp.* **2016**, *109*, e53069. [CrossRef] [PubMed]

19. Pfeiffer, E.; Kegel, V.; Zeilinger, K.; Hengstler, J.G.; Nüssler, A.K.; Seehofer, D.; Damm, G. Featured Article: Isolation, characterization, and cultivation of human hepatocytes and non-parenchymal liver cells. *Exp. Biol. Med.* **2015**, *240*, 645–656. [CrossRef] [PubMed]

20. Freyer, N.; Knöspel, F.; Strahl, N.; Amini, L.; Schrade, P.; Bachmann, S.; Damm, G.; Seehofer, D.; Jacobs, F.; Monshouwer, M.; et al. Hepatic differentiation of human induced pluripotent stem cells in a perfused three-dimensional multicompartment bioreactor. *BioRes. Open Access* **2016**, *5*, 235–248. [CrossRef] [PubMed]

21. Livak, K.J.; Schmittgen, T.D. Analysis of relative gene expression data using real-time quantitative PCR and the $2^{-\Delta\Delta Ct}$ Method. *Methods* **2001**, *25*, 402–408. [CrossRef] [PubMed]

22. Hinson, J.A.; Roberts, D.W.; James, L.P. Mechanisms of acetaminophen-induced liver necrosis. *Handb. Exp. Pharmacol.* **2010**, *196*, 369–405. [CrossRef]

23. Treyer, A.; Müsch, A. Hepatocyte polarity. *Compr. Physiol.* **2013**, *3*, 243–287. [CrossRef] [PubMed]

24. Tormos, A.M.; Taléns-Visconti, R.; Bonora-Centelles, A.; Pérez, S.; Sastre, J. Oxidative stress triggers cytokinesis failure in hepatocytes upon isolation. *Free Radic. Res.* **2015**, *49*, 927–934. [CrossRef] [PubMed]

25. Xie, Y.; McGill, M.R.; Dorko, K.; Kumer, S.C.; Schmitt, T.M.; Forster, J.; Jaeschke, H. Mechanisms of acetaminophen-induced cell death in primary human hepatocytes. *Toxicol. Appl. Pharmacol.* **2014**, *279*, 266–274. [CrossRef] [PubMed]

26. Prill, S.; Jaeger, M.S.; Duschl, C. Long-term microfluidic glucose and lactate monitoring in hepatic cell culture. *Biomicrofluidics* **2014**, *8*, 034102. [CrossRef] [PubMed]

27. Cummings, B.S.; Wills, L.P.; Schnellmann, R.G. Measurement of cell death in mammalian cells. *Curr. Protoc. Pharmacol.* **2012**. [CrossRef]

28. Silva, M.T. Secondary necrosis: The natural outcome of the complete apoptotic program. *FEBS Lett.* **2010**, *584*, 4491–4499. [CrossRef] [PubMed]

29. Jialal, I.; Sokoll, L.J. Clinical utility of lactate dehydrogenase: A historical perspective. *Am. J. Clin. Pathol.* **2015**, *143*, 158–159. [CrossRef] [PubMed]

30. Cassidy, W.M.; Reynolds, T.B. Serum lactic dehydrogenase in the differential diagnosis of acute hepatocellular injury. *J. Clin. Gastroenterol.* **1994**, *19*, 118–121. [CrossRef] [PubMed]

31. Wolf, P.L. Biochemical diagnosis of liver disease. *Indian J. Clin. Biochem.* **1999**, *14*, 59–90. [CrossRef] [PubMed]

32. Choi, K.; Pfund, W.P.; Andersen, M.E.; Thomas, R.S.; Clewell, H.J.; LeCluyse, E.L. Development of 3D dynamic flow model of human liver and its application to prediction of metabolic clearance of 7-ethoxycoumarin. *Tissue Eng. Part C Methods* **2014**, *20*, 641–651. [CrossRef] [PubMed]

33. Ullrich, A.; Stolz, D.B.; Ellis, E.C.; Strom, S.C.; Michalopoulos, G.K.; Hengstler, J.G.; Runge, D. Long term cultures of primary human hepatocytes as an alternative to drugtesting in animals. *ALTEX* **2009**, *26*, 295–302. [CrossRef] [PubMed]

34. Bolleyn, J.; Rogiers, V.; Vanhaecke, T. Functionality testing of primary hepatocytes in culture by measuring urea synthesis. *Methods Mol. Biol.* **2015**, *1250*, 317–321. [CrossRef] [PubMed]

35. Simmons, D.L.; Botting, R.M.; Hla, T. Cyclooxygenase isozymes: The biology of prostaglandin synthesis and inhibition. *Pharmacol. Rev.* **2004**, *56*, 387–437. [CrossRef] [PubMed]

36. Norris, C.A.; He, M.; Kang, L.I.; Ding, M.Q.; Radder, J.E.; Haynes, M.M.; Yang, Y.; Paranjpe, S.; Bowen, W.C.; Orr, A.; et al. Synthesis of IL-6 by hepatocytes is a normal response to common hepatic stimuli. *PLoS ONE* **2014**, *9*, e96053. [CrossRef] [PubMed]

37. Kalinski, P. Regulation of immune responses by prostaglandin E2. *J. Immunol.* **2012**, *188*, 21–28. [CrossRef] [PubMed]

38. Manyike, P.T.; Kharasch, E.D.; Kalhorn, T.F.; Slattery, J.T. Contribution of CYP2E1 and CYP3A to acetaminophen reactive metabolite formation. *Clin. Pharmacol. Ther.* **2000**, *67*, 275–282. [CrossRef] [PubMed]

39. Lee, S.S.; Buters, J.T.; Pineau, T.; Fernandez-Salguero, P.; Gonzalez, F.J. Role of CYP2E1 in the hepatotoxicity of acetaminophen. *J. Biol. Chem.* **1996**, *271*, 12063–12067. [CrossRef] [PubMed]

40. Patten, C.J.; Thomas, P.E.; Guy, R.L.; Lee, M.; Gonzalez, F.J.; Guengerich, F.P.; Yang, C.S. Cytochrome P450 enzymes involved in acetaminophen activation by rat and human liver microsomes and their kinetics. *Chem. Res. Toxicol.* **1993**, *6*, 511–518. [CrossRef] [PubMed]

41. Raucy, J.L.; Lasker, J.M.; Lieber, C.S.; Black, M. Acetaminophen activation by human liver cytochromes P450IIE1 and P450IA2. *Arch. Biochem. Biophys.* **1989**, *271*, 270–283. [CrossRef]

42. Laine, J.E.; Auriola, S.; Pasanen, M.; Juvonen, R.O. Acetaminophen bioactivation by human cytochrome P450 enzymes and animal microsomes. *Xenobiotica* **2009**, *39*, 11–21. [CrossRef] [PubMed]

43. Sarich, T.; Kalhorn, T.; Magee, S.; Al-Sayegh, F.; Adams, S.; Slattery, J.; Goldstein, J.; Nelson, S.; Wright, J. The effect of omeprazole pretreatment on acetaminophen metabolism in rapid and slow metabolizers of S-mephenytoin. *Clin. Pharmacol. Ther.* **1997**, *62*, 21–28. [CrossRef]

44. Khosravi, M.; Saadat, I.; Karimi, M.H.; Geramizadeh, B.; Saadat, M. Glutathione S-transferase omega 2 genetic polymorphism and risk of hepatic failure that lead to liver transplantation in iranian population. *Int. J. Organ Transplant. Med.* **2013**, *4*, 16–20. [PubMed]

45. Chen, C.; Hennig, G.E.; Manautou, J.E. Hepatobiliary excretion of acetaminophen glutathione conjugate and its derivatives in transport-deficient (TR-) hyperbilirubinemic rats. *Drug Metab. Dispos.* **2003**, *31*, 798–804. [CrossRef] [PubMed]

46. Wang, J.X.; Zhang, C.; Fu, L.; Zhang, D.G.; Wang, B.W.; Zhang, Z.H.; Chen, Y.H.; Lu, Y.; Chen, X.; Xu, D.X. Protective effect of rosiglitazone against acetaminophen-induced acute liver injury is associated with down-regulation of hepatic NADPH oxidases. *Toxicol. Lett.* **2017**, *265*, 38–46. [CrossRef] [PubMed]

47. Au, S.H.; Chamberlain, M.D.; Mahesh, S.; Sefton, M.V.; Wheeler, A.R. Hepatic organoids for microfluidic drug screening. *Lab Chip* **2014**, *14*, 3290–3299. [CrossRef] [PubMed]

48. Rinaldi, C.; Grunseich, C.; Sevrioukova, I.F.; Schindler, A.; Horkayne-Szakaly, I.; Lamperti, C.; Landouré, G.; Kennerson, M.L.; Burnett, B.G.; Bönnemann, C.; et al. Cowchock syndrome is associated with a mutation in apoptosis-inducing factor. *Am. J. Hum. Genet.* **2012**, *91*, 1095–1102. [CrossRef] [PubMed]

49. Guicciardi, M.E.; Malhi, H.; Mott, J.L.; Gores, G.J. Apoptosis and necrosis in the liver. *Compr. Physiol.* **2013**, *3*, 977–1010. [CrossRef] [PubMed]

50. Jaeschke, H.; Bajt, M.L. Intracellular signaling mechanisms of acetaminophen-induced liver cell death. *Toxicol. Sci.* **2006**, *89*, 31–41. [CrossRef] [PubMed]

51. Zamek-Gliszczynski, M.J.; Hoffmaster, K.A.; Tian, X.; Zhao, R.; Polli, J.W.; Humphreys, J.E.; Webster, L.O.; Bridges, A.S.; Kalvass, J.C.; Brouwer, K.L. Multiple mechanisms are involved in the biliary excretion of acetaminophen sulfate in the rat: Role of MRP2 and BCRP1. *Drug Metab. Dispos.* **2005**, *33*, 1158–1165. [CrossRef] [PubMed]

52. Godoy, P.; Hewitt, N.J.; Albrecht, U.; Andersen, M.E.; Ansari, N.; Bhattacharya, S.; Bode, J.G.; Bolleyn, J.; Borner, C.; Böttger, J.; et al. Recent advances in 2D and 3D in vitro systems using primary hepatocytes, alternative hepatocyte sources and non-parenchymal liver cells and their use in investigating mechanisms of hepatotoxicity, cell signaling and ADME. *Arch. Toxicol.* **2013**, *87*, 1315–1530. [CrossRef] [PubMed]

53. Bhise, N.S.; Manoharan, V.; Massa, S.; Tamayol, A.; Ghaderi, M.; Miscuglio, M.; Lang, Q.; Shrike Zhang, Y.; Shin, S.R.; Calzone, G.; et al. A liver-on-a-chip platform with bioprinted hepatic spheroids. *Biofabrication* **2016**, *8*, 014101. [CrossRef] [PubMed]

54. Ma, C.; Zhao, L.; Zhou, E.M.; Xu, J.; Shen, S.; Wang, J. On-chip construction of liver lobule-like microtissue and its application for adverse drug reaction assay. *Anal. Chem.* **2016**, *88*, 1719–1727. [CrossRef] [PubMed]

55. Cairney, D.G.; Beckwith, H.K.; Al-Hourani, K.; Eddleston, M.; Bateman, D.N.; Dear, J.W. Plasma paracetamol concentration at hospital presentation has a dose-dependent relationship with liver injury despite prompt treatment with intravenous acetylcysteine. *Clin. Toxicol.* **2016**, *54*, 405–410. [CrossRef] [PubMed]

56. Whitcomb, D.C.; Block, G.D. Association of acetaminophen hepatotoxicity with fasting and ethanol use. *JAMA* **1994**, *272*, 1845–1850. [CrossRef] [PubMed]

57. Jaeschke, H.; Williams, C.D.; Ramachandran, A.; Bajt, M.L. Acetaminophen hepatotoxicity and repair: The role of sterile inflammation and innate immunity. *Liver Int.* **2012**, *32*, 8–20. [CrossRef] [PubMed]

58. McCuskey, R.S. Sinusoidal endothelial cells as an early target for hepatic toxicants. *Clin. Hemorheol. Microcirc.* **2006**, *34*, 5–10. [PubMed]

59. Kegel, V.; Pfeiffer, E.; Burkhardt, B.; Liu, J.L.; Zeilinger, K.; Nüssler, A.K.; Seehofer, D.; Damm, G. Subtoxic concentrations of hepatotoxic drugs lead to Kupffer cell activation in a human in vitro liver model: An approach to study DILI. *Mediators Inflamm.* **2015**, *2015*, 640631. [CrossRef] [PubMed]

60. Prodanov, L.; Jindal, R.; Bale, S.S.; Hegde, M.; McCarty, W.J.; Golberg, I.; Bhushan, A.; Yarmush, M.L.; Usta, O.B. Long-term maintenance of a microfluidic 3D human liver sinusoid. *Biotechnol. Bioeng.* **2016**, *113*, 241–246. [CrossRef] [PubMed]

61. Rennert, K.; Steinborn, S.; Gröger, M.; Ungerböck, B.; Jank, A.M.; Ehgartner, J.; Nietzsche, S.; Dinger, J.; Kiehntopf, M.; Funke, H.; et al. A microfluidically perfused three dimensional human liver model. *Biomaterials* **2015**, *71*, 119–131. [CrossRef] [PubMed]

62. Allen, J.W.; Khetani, S.R.; Bhatia, S.N. In vitro zonation and toxicity in a hepatocyte bioreactor. *Toxicol. Sci.* **2005**, *84*, 110–119. [CrossRef] [PubMed]

bioengineering

MDPI

Article

A Cardiac Cell Outgrowth Assay for Evaluating Drug Compounds Using a Cardiac Spheroid-on-a-Chip Device

Jonas Christoffersson [1], Florian Meier [2], Henning Kempf [3], Kristin Schwanke [3], Michelle Coffee [3], Mario Beilmann [2], Robert Zweigerdt [3],* and Carl-Fredrik Mandenius [1],*

[1]   Division of Biotechnology, Department of Physics, Chemistry and Biology (IFM), Linköping University,
      58183 Linköping, Sweden; jonas.christoffersson@liu.se
[2]   Boehringer Ingelheim Pharma GmbH and Co. KG, Nonclinical Drug Safety Germany,
      D-88397 Biberach an der Riss, Germany; florian.meier@boehringer-ingelheim.com (F.M.);
      mario.beilmann@boehringer-ingelheim.com (M.B.)
[3]   Leibniz Research Laboratories for Biotechnology and Artificial Organs (LEBAO), Hannover Medical School,
      Carl-Neuberg-Str. 1, 30625 Hannover, Germany; Kempf.Henning@mh-hannover.de (H.K.);
      Schwanke.Kristin@mh-hannover.de (K.S.); Coffee.Michelle@mh-hannover.de (M.C.)
*   Correspondence: Zweigerdt.Robert@mh-hannover.de (R.Z.); carl-fredrik.mandenius@liu.se (C.-F.M.);
      Tel.: +49-511-5328773 (R.Z.); +46-013-288967 (C.-F.M.)

Received: 9 March 2018; Accepted: 1 May 2018; Published: 4 May 2018

Abstract: Three-dimensional (3D) models with cells arranged in clusters or spheroids have emerged as valuable tools to improve physiological relevance in drug screening. One of the challenges with cells cultured in 3D, especially for high-throughput applications, is to quickly and non-invasively assess the cellular state in vitro. In this article, we show that the number of cells growing out from human induced pluripotent stem cell (hiPSC)-derived cardiac spheroids can be quantified to serve as an indicator of a drug's effect on spheroids captured in a microfluidic device. Combining this spheroid-on-a-chip with confocal high content imaging reveals easily accessible, quantitative outgrowth data. We found that effects on outgrowing cell numbers correlate to the concentrations of relevant pharmacological compounds and could thus serve as a practical readout to monitor drug effects. Here, we demonstrate the potential of this semi-high-throughput "cardiac cell outgrowth assay" with six compounds at three concentrations applied to spheroids for 48 h. The image-based readout complements end-point assays or may be used as a non-invasive assay for quality control during long-term culture.

Keywords: 3D cell culture; microfluidics; organ-on-a-chip; cardiac spheroids; cardiomyocytes; induced pluripotent stem cells (iPSCs); drug screening

---

1. Introduction

The recent development of perfused three-dimensional (3D) cell culture models, or organs-on-chip, offers the possibility to investigate biological responses of chemicals and pharmaceuticals in a model that better mimics the in vivo cell environment than conventional two-dimensional culture models [1,2]. Therefore, results from such assays are believed to increase the predictivity of drug effects on human tissue such as efficacy and toxicity. Advanced in vitro assays may thus better predict harmful or ineffective chemicals before they enter the long and expensive drug development process. Common approaches to create a 3D cell environment are to embed the cells in a hydrogel matrix such as collagen [3] or Matrigel [4], or to let the cells aggregate into cell spheroids [5–7]. A critical challenge for both 2D and 3D-based assays is to examine the impact of compounds on the target cells without substantial interference. For continuous non-invasive assaying, several methods have been developed to analyze the supernatant of the cell culture medium to reveal the cellular state in sequential off-line monitoring of biomarkers [8,9]. Furthermore, for cardiac cells, standard methods include the recording of beating frequency and electrocardiographic recording using microelectrode arrays which can be performed non-invasively [10,11]. However, recording videos of cells is time consuming, and electrocardiography is mostly performed on 2D cardiomyocytes. Analysis of cell growth and morphology have previously been reported for several cell types such as neurites in the neuronal network formation assay and endothelial cells in the wound healing assay [12,13]. However, with respect to cardiac assays, the outgrowth of cells has been described as a naturally occurring process which, in primary tissue, may result from cardiac progenitor cells [14]. Compared to conventional static conditions, dynamic cell cultures have been shown to have positive effects on several cell types [15–17] and also to support functional outputs of cardiac aggregates [18].

In this article, we combine recent progress on the derivation of human pluripotent stem cell-cardiomyocytes (CMs), their use for engineering cardiac tissue including spheroids, and in microfluidics technology for developing novel drug testing assays. The approach is based on quantifying the number of cells growing out from cardiac spheroids within a defined time and area, by combining solvent controls versus exposure to six compounds at three concentrations. Non-invasive, microscopy-based assessment showed substantial effects of doxorubicin, endothelin-1 (both decreasing cell outgrowth), and amiodarone (support cell outgrowth). To objectively determine the cell outgrowth around the spheroids, cell nuclei were stained and counted using a high content imaging system which also revealed the effect of phenylephrine (increased outgrowth). Comparisons were also made between static and dynamic cultures, and between cardiac spheroids derived from two different human induced pluripotent stem cell (hiPSC) lines, both confirming the drug- and dose-dependent effects.

With the challenges of analyzing 3D cell spheroids in mind, this novel approach for investigating the effect of chemicals or drug compounds could be used as a compliment to invasive end-point assays or as a non-invasive quality control tool used during long-term cultures.

2. Materials and Methods

2.1. Cell Lines and Preparation of Cardiac Spheroids

Cardiac spheroids, each consisting of approximately 2500 cells (~250 µm in diameter) were generated as follows. The human induced pluripotent stem cell lines (hiPSC) SFC086-03-01 and SFC840-03-01 (referred to as SFC086 and SFC840, respectively, and derived by the StemBANCC initiative [19] http://stembancc.org/; received from the Human Biomaterials Resource Centre, University of Birmingham (http://www.birmingham.ac.uk/facilities/hbrc)) were cultured and differentiated by recently established protocols in suspension culture [20–22] to achieve a cardiomyocyte (CM) content of ~90% (SFC086) and >90% (SFC840), respectively (Figure 1). Briefly, cells were dissociated using collagenase IV (Life Technologies) and subsequently resuspended in medium consisting of Iscove's modified Dulbecco's medium with GlutaMAX™ (Life Technologies/ Thermo Fisher Scientific, Waltham, MA, USA) supplemented with 20% fetal bovine serum, 0.2 mM L-glutamine, 0.1 mM b-mercaptoethanol, 1% non-essential amino acids (*v/v*), 1 mg/mL penicillin, and 1 U/mL streptomycin and 10 µM Rho-associated coiled-coil kinase (ROCK) inhibitor Y-27632. Cells were then reseeded into Statarrays© MCA96-16.224-PSLA Low Attachment Surface plates (300MICRONS, Karlsruhe, Germany) with the concentration of 2500 cells per microcavity, in 300 µL of the same medium to form spheroids within 48 h. Spheroids were recovered by pipetting gently, transferred to RB+ medium consisting of RPMI1640 supplemented with B27 with insulin (Life Technologies) and used for experiments within 5–10 days after generation.

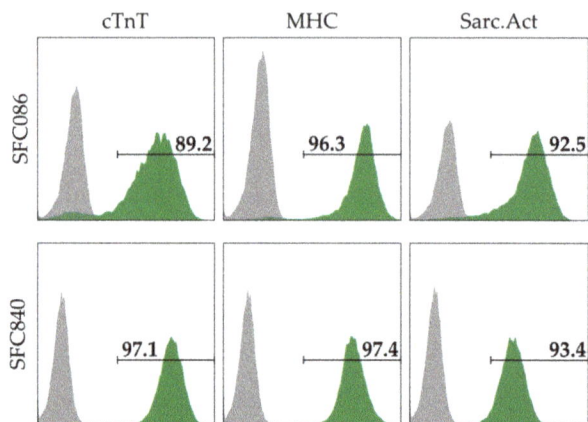

Figure 1. Flow cytometry-based assessment of cardiomyocyte (CM) content using immunofluorescent stains specific to cardiac troponin T (cTnT), pan-myosin heavy chain (MHC), and sarcomeric actinin (Sarc.Act). Depending on the marker, the CM content of the SFC086 cell line and the SFC840 cell line was ~89–96% and ~93–97% respectively. CM content was assessed at day 10–14 after induction of differentiation, i.e., when the cells were used for spheroid formation.

2.2. Flow Cytometry and Immunofluorescent Staining

For intracellular staining, 1.5×10^5 differentiated cells were fixed/permeabilized according to manufacturer's instructions (Fix&Perm-kit by An der Grub, Thermo Fisher Scientific, Waltham, MA, USA). Antibodies specific to cardiac troponin T (1:200, clone 13-11, Thermo Fisher Scientific), sarcomeric actinin (α-ACTININ; 1:800, clone EA-53, Sigma-Aldrich, St. Louis, MO, USA), myosin heavy chain (MHC; 1:25, Hybridoma Bank, Iowa City, IA, USA; 1:2000, clone NOQ7.5.4D, Sigma-Aldrich), NKX2.5 (1:200; clone H-114; Santa Cruz, California, CA, USA), and respective isotype controls (Dako, Agilent, Santa Clara, CA, USA) were detected using appropriate Cy3-/Cy5-conjugated antibodies (1:200; Jackson Immunoresearch Laboratories, West Grove, PA, USA). After washing, signals were detected using Cy3-labeled donkey anti-mouse IgM (1:200; Jackson Immunoresearch Laboratories) on the Accuri C6 flow cytometer (BD Biosciences, Franklin Lakes, NJ, USA). Data were analyzed using FlowJo (Treestar, FlowJo, LLC, Ashland, OR, USA).

Plated aggregates/outgrowing cells were fixed with 4% paraformaldehyde, 15 min, room temperature (RT). After blocking by Tris-buffered saline (5% donkey serum, 0.25% Triton X-100), cells were incubated with primary/secondary antibodies listed above, respectively. Nuclei were DAPI-stained, and samples were analyzed using the Axio Observer A1 (Zeiss, Jena, Germany) or a DM IRB/TCS SP2 confocal microscope system (Leica, Wetzlar, Germany).

2.3. Microfluidic System and Experimental Procedure for Compound Testing

Microfluidic channel slides with dimensions $17 \times 3.8 \times 0.4$ mm (length × width × height) (Ibidi μSlide VI$^{0.4}$, untreated) were coated to facilitate cell attachment by adding 30 μL laminin (100 μg/mL) to each channel for 1 h at room temperature, followed by washing with PBS and cell culture medium before seeding the cell spheroids. Cardiac spheroids were infused by adding 30 μL with approximately 20 spheroids to one of the wells and rapidly removing 20 μL cell culture medium from the opposite well. The spheroids were allowed to attach to the surface of the channel at 37 °C and 5% CO_2. After 16 h incubation, the cell culture medium in each channel was removed and substituted to 90 μL compound supplemented cell culture medium or control (0.25% DMSO) before mounting the slide on a motorized rocking table for continuous alternating perfusion of cell culture medium with a frequency of 0.5 Hz at 37 °C and 5% CO_2. (see also Supplementary Figure S1). Applied compounds

Doxorubicin, endothelin-1, acetylsalicylic acid, isoproterenol, phenylephrine, and amiodarone were all purchased from Sigma and diluted in a final concentration of 0.25% DMSO in cell culture medium. The concentrations used were 0.04 nM, 1.11 nM, and 10 nM of endothelin-1, and 0.04 μM, 1.11 μM, and 10 μM for all other compounds. After 48 h of treatment, spheroids were washed with PBS, fixed in 4% PFA for 1 h at room temperature, and the cell nuclei were then stained with Hoechst 33342.

2.4. Image Processing and Nuclei Quantification

Images of the spheroids were captured by a camera (Canon PowerShot A640, Canon Tokyo, Japan) attached to a phase microscope (Zeiss Axiovert 40C, Zeiss) or, for quantification of the number of nuclei around the spheroids, by a high content imaging system (Opera Phenix, Perkin Elmer, Waltham, MA, USA). Images were taken with a 5× objective in confocal mode. Image analysis was performed with the Harmony (Perkin Elmer) software. The number of nuclei was normalized to the area of analysis around each spheroid to account for variations in size (see Supplementary Figure S2). Statistical difference between control samples of the SFC086 and SFC840 cell lines, both run in triplicates and counting the number of nuclei around at least 18 spheroids in each channel, was determined by a two-pair *t*-test at a level of $\alpha = 0.05$. Statistical difference between control samples and the number of nuclei around at least six spheroids for each compound treatment was determined by a z-test at a level of $\alpha = 0.05$.

3. Results

During routine culture of cardiac spheroids on laminin coated surfaces, we observed that some cardiac cells tended to grow out from the aggregates and attach to the surrounding surface. From this observation we hypothesized that this effect could be exploited for developing a cell-based assay able to assess effects of a drug (Figure 2). Specifically, the assay should be based on the assumption that exposing cardiac spheroids to compounds would affect the number of surrounding cells, and by that, assess both negative (decreased cell number) and positive (increased cell number) effects on the particular cell type's ability to proliferate or migrate, in relation to the compound.

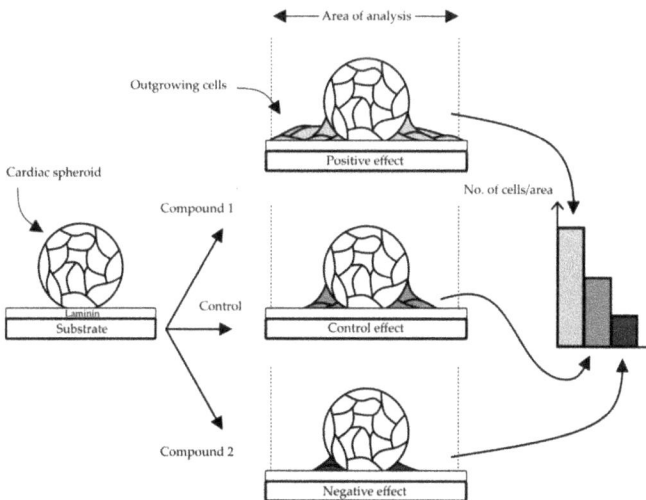

Figure 2. Concept of the proposed cardiac cell outgrowth assay. An outgrowth of cells from spheroids over time was observed on laminin coated surfaces (control) and the number of cells within a defined area should be possible to quantify and potentially reveal an increased (positive) or decreased (negative) effect depending on the compound.

3.1. Design of the Cell Outgrowth Assay

To realize the outgrowth assay in an appropriate microfluidic format, the design had to accommodate several critical functions. These included a contained fluidized space for the cell spheroids, a geometric format suitable for high content screening, a device with several parallel cell culture areas to allow a semi-high throughput procedure, and an expedient way to perfuse the device with cell culture medium. These criteria were matched by an existing commercial device on a microscope slide format (Figure 3). The device consisted of six parallel channels with adjacent reservoirs that could be fixed on a motorized rocker to drive the perfusion of cell culture medium by gravity. By coating the channels with laminin, it was possible to capture the cardiac spheroids in the device for subsequent imaging by light microscopy and confocal imaging. The cardiac spheroids in each image were identified by automatic thresholding by the high content imaging software. A contour was applied 12 μm (5 pixels) outside the boundary of the spheroids to create a space from the dense cardiac spheroids. A second contour was applied 117 μm (50 pixels) from the spheroids to define the area of analysis within the two contours. Because several areas were spotted where closely neighboring spheroids occasionally fused into larger clusters, the area of analysis was made to ensure distinctly separated clusters. The labelling of cardiac specific markers on cells growing out from the spheroids confirmed the predominant presence of CMs (Figure 4) and a relatively small portion of non-cardiac cells (white arrows, Figure 4) in line with the flow cytometry data, suggesting high content of CMs in the differentiated cell suspension used for spheroid formation.

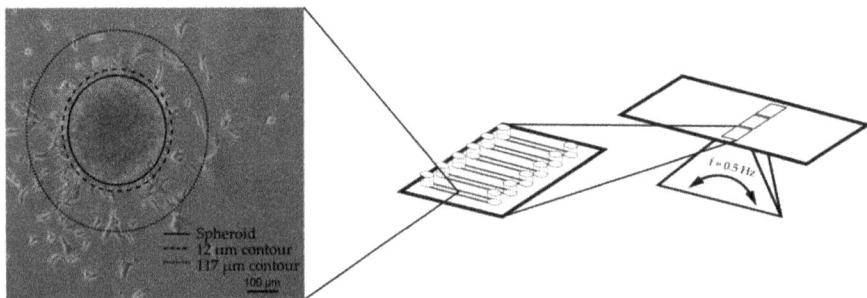

Figure 3. Cardiac spheroids were seeded in fluidic channels on a microscope slide-format and fixed on a rocking platform to drive the perfusion through the device by gravity. The number of nuclei within the area confined by the 12 μm and 117 μm contours were counted by high content imaging.

3.2. Verification of the Assay Performance with Six Compounds

Six compounds, all with either potential negative, negligible, or positive effect on cardiac cell growth, were chosen for verifying the assay performance (Table 1). The compounds included doxorubicin, a drug compound used in chemotherapy with well-documented toxic effects on heart cells [23]; endothelin-1, an established vasoconstrictor and hypertrophy inducer [24]; acetylsalicylic acid, an analgesic and anti-inflammatory compound [25]; isoproterenol, a β-adrenergic receptor agonist and positive inotrope [26]; phenylephrine, a α-adrenergic receptor agonist and vasoconstrictor [27]; and amiodarone, a K+ channel blocker and antiarrhythmic agent [28]. These compounds were selected due to their diverse but well-established effects on cardiac cells (see Table 1). We assume that these drugs showed low and similar interactions with the laminin-coated surfaces of the device, although it cannot be excluded that minor outgrowth effects might have occurred in the experiments.

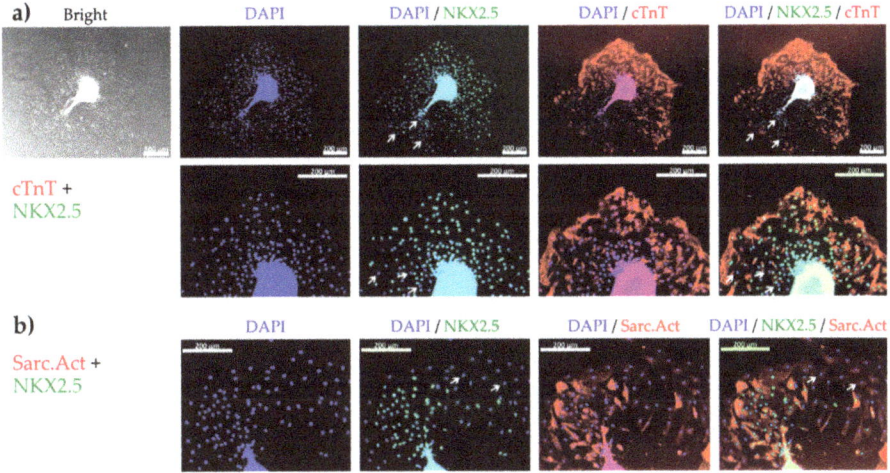

Figure 4. The presence of cardiomyocytes (CM) among cells growing out from the spheroids 48 h after plating determined by fluorescence microscopy of CM specific markers. (**a**) Bright field image of cell outgrowth from a spheroid and subsequent fluorescence microscopy showing nuclei (DAPI, blue), early CM specific transcription factor (NKX2.5, green), and cardiac troponin T (cTnT, red), at 10× (top) and 20× (bottom) magnification. (**b**) Fluorescence microscopy showing nuclei (blue), NKX2.5 (green), and sarcomeric actinin (Sarc.Act, red) at 20× magnification. Most of the cells were positive for both NKX2.5 and cTnT or NKX2.5 and Sarc.Act revealing their CM phenotype. White arrows show NKX2.5 negative nuclei indicating the presence of non-cardiac cells. Scale bars represent 200 μm.

Table 1. The compounds used in these experiments and their expected effect on cell outgrowth.

Compound	Type	Mechanism	Expected Effect (−/\/+) [1]
Doxorubicin	Chemotherapeutic	DNA intercalator	−
Endothelin-1	Vasoconstrictor	Hypertrophy inducer	−
Acetylsalicylic acid	Analgesic	Anti-inflammation	\
Isoprenterenol	Positive inotrope	β-adrenergic receptor agonist	+
Phenylephrine	Vasoconstrictor	α-adrenergic receptor agonist	+
Amiodarone	Antiarrhythmic agent	K$^+$ channel blocker	+

[1] The expected effect was either negative (−), negligible (), or positive (+) with regard to cell number increase in the area of analyses.

After 48 h of treatment, cells growing out from the cardiac spheroid were first visualized by phase contrast microscopy (Figure 5). All spheroids were still contractile at the time of fixation, but the rate was not quantified during these experiments. In control samples, cells were found inhabiting the laminin coated surface surrounding the spheroids. A drastic decrease of cell outgrowth was apparent in the presence of doxorubicin at 1.11 μM and 10 μM, and of endothelin-1 at 1.11 nM and 10 nM. The opposite effect was observed with amiodarone at 1.11 μM and 10 μM inducing increase cell outgrowth. In the presence of acetylsalicylic acid, isoproterenol, and phenylephrine, minor effects compared to control sample were observed. For more accurate quantification, cell nuclei stained with Hoescht 33342 were assessed and counted within the area of analysis by the Opera high content imaging system (Figure 6a). As presented in Figure 6b, the effect of the highest doses of doxorubicin, endothelin-1, and amiodarone, which were readily indicated by more qualitative phase contrast monitoring, was confirmed. Furthermore, a significant difference compared to control samples was now detected already at 0.04 μM of doxorubicin, and at 1.11 μM and 10 μM of phenylephrine.

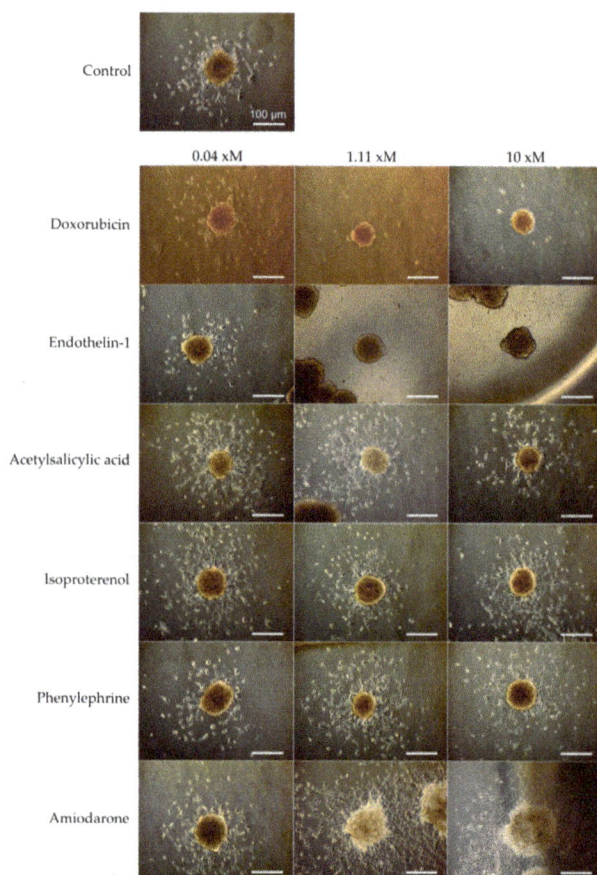

Figure 5. Micrographs of cardiac spheroids (SFC086) showing the outgrowth of cells from the aggregates and the effect of six drug compounds at three concentrations after 48 h on the number of cells around the spheroids. For endothelin-1, x = n (nM). For all other compounds, x = μ (μM). Scale bar represents 100 μm.

Indications of more sensitive cells under static conditions, when compared to the dynamic control, were noted, as fewer nuclei per area surrounded the spheroids when treated with enothelin-1 from 0.04 nM and in acetylsalicylic acid at 1.11 μM (Figure 6c). However, any clear differences were neither found for endothelin-1 nor for acetylsalicylic acid but could be of interest to study in a larger experiment. Furthermore, as compound-exposed cardiac spheroids at static conditions were compared to control samples at dynamic conditions, any distinct conclusions would be difficult to draw from these experiments. Nevertheless, similar patterns, with fewer cells around endothelin-1 exposed spheroids compared to acetylsalicylic acid exposed spheroids, were observed at both dynamic and static conditions.

Finally, we compared cardiac spheroids derived from the hiPSC lines SFC086 versus SFC840 (cardiomyocyte content of ~90–95% for both lines) in our assay. Results were highly consistent between both lines (Figure 6d), suggesting cell line independent reproducibility and general relevance of the method.

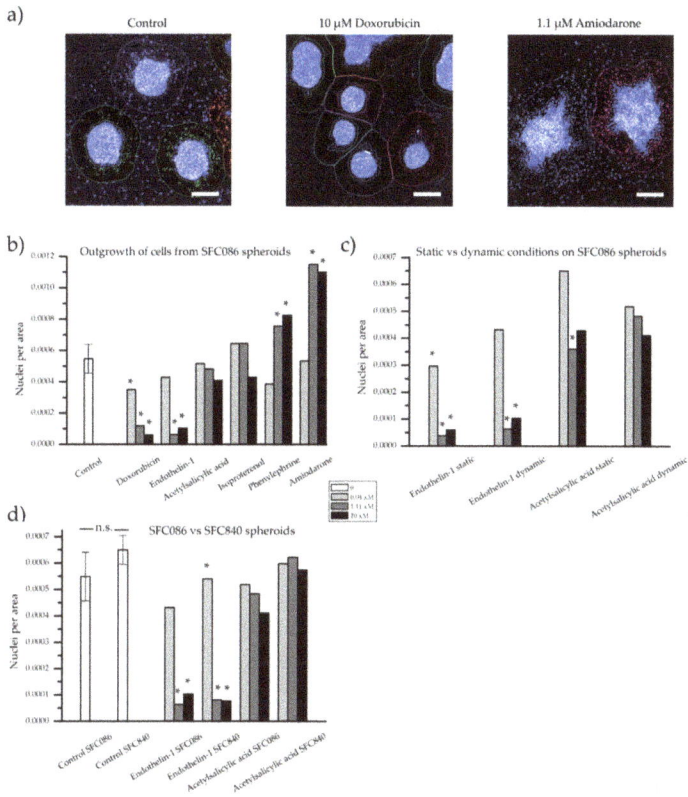

Figure 6. Quantification of the number of nuclei visible outside a defined area of the spheroids. (**a**) Examples of the applied contours around spheroids stained with Hoechst 33342 and captured by high content imaging. Number of nuclei per area of analysis from (**b**) the SFC086 cell line under dynamic conditions exposed to six compounds at three concentrations, (**c**) the SFC086 cell line at static and dynamic conditions exposed to two compounds, and (**d**) the SFC086 cell line compared to the SFC840 cell line when exposed to two compounds. Control samples are presented as mean ± standard deviation of three experiments, * denotes a significant difference of the sample from the dynamic control. No significance (n.s.) was observed between the two control samples (SFC086 and SFC840). For endothelin-1, x = n (nM). For all other compounds, x = μ (μM). Scalebar represents 200 μm.

4. Discussion

The results demonstrated a pragmatic quantitative strategy to assess drug effect on cardiac cells at the interplay of 3D/2D and static/dynamic culture. For this to occur, a method for capturing cell spheroids in a fluidic device followed by continuous dynamic culture was established. The setup is robust and simple to use even for newcomers in the field of microfluidic cell culture users. We show experimental conditions enabling to use the "cardiac cell outgrowth from 3D spheroids" as a quantitative measurement to examine significant drug effects. It may be argued that the assay not completely mirrors 3D conditions since the assay partly takes place outside the spheroids and should therefore be considered a "2.5D" assay. However, significant drug interactions representative for the assay readout occur inside the spheroids.

The method is fast and enables continuous non-invasive cell monitoring by standard light microscopy as a readout, which deems compatible with automated high content imaging and screening.

Subsequently, the approach can be used as an end-point assay in combination with a potentially more accurately quantifiable fluorescence-based analysis of counterstained nuclei.

To validate assays' utility for monitoring drug effects, we decided to use five compounds with known impact on the heart (doxorubicin, endothelin-1, isoproterenol, phenylephrine, and amiodarone), and one control drug that may not affect cardiomyocyte function (acetylsalicylic acid), all tested at three concentrations. The decreased number of cells around spheroids exposed to doxorubicin can be attributed to the compound's established cytotoxicity, which has previously been shown to cause cardiac cell death [23]. Endothelin-1 is associated with hypertrophic responses on cardiac cells such as increased cell size and elevated levels of B-type natriuretic peptide (BNP) expression [19], myofibrillar disarray [29], and has been shown to increase the survival of doxorubicin treated rat cardiomyocytes in short-term (24 h) [30]. These data suggest that cytotoxicity may not be predominantly responsible for the decreased cell outgrowth in spheroids exposed to endothelin-1. However, the reported myofibrillar disarray due to ET-1 [29] may suggest an overall detrimental impact of the drug on cytoskeletal function, thereby reducing cardiomyocytes motility and thus limiting their outgrowth in our assay. Whilst further mechanistic analysis is required, these observations highlight the sensitivity of our assay since the effect was readily observed at a relative low ET-1 dose of 1.11 nM.

No significant effect could be observed at any concentration of isoproterenol. The beta-adrenergic agonist isoproterenol is therapeutically applied to increase the heart rate in patients suffering from bradycardia (slow heart rate). In our approach, we have solely focused on differences in cell morphological outputs and not measured functional attributes such as the beating rate, but notably the general experimental setup is compatible with beating rate assessment.

Equivalent to endothelin-1, phenylephrine is a known vasoconstrictor [27]. Interestingly, an opposite effect of these two drugs, that is an increased outgrowth upon phenylephrine administration versus outgrowth suppression by ET-1, was observed. Phenylephrine mimics epinephrine which both have been shown to promote cell growth of cardiac fibroblasts and increasing the cell size of neonatal cardiomyocytes [31].

The increased number of cells around spheroids treated with amiodarone is less predictive from published literature. Previous studies on hiPSC-derived cardiomyocytes showed contradicting effects on cells' beating rate in response to amiodarone ranging from no effect at up to 10 µM [32] to decreased beating at up to 100 µM [33] and increased frequency at up to 100 µM [34]; this heterogeneity may be attributed to cell- or assay-dependent effects. However, we did not find published data on CM proliferation or motility in response to amiodarone administration, which has apparently not been identified by other assays before.

Besides the use in advanced drug testing assays, human pluripotent stem cell-derived CMs have also been considered in regenerative medicine [35]. One central problem in that field is the low implantation efficiency and pure cell distribution in the heart post cell administration [36–38]. Although we have not performed animal studies here, it is tempting to speculate that those drugs, that support CMs outgrowth in our in vitro assay, may also support proliferation and better distribution of transplanted cells in cardiac tissue, thereby facilitating heart repair.

The CMs used for spheroid generation were derived by well-defined procedures in suspension culture [20,21] using chemical Wnt pathway modulators to ensure lineage-specific differentiation and high CMs content [22,39]. It is worth noting that such CMs represent an immature phenotype with respect to their gene and protein expression pattern and physiological properties [40,41]. Compared to their functional counterparts in the adult heart, which are known to be cell cycle arrested and non-proliferative, human pluripotent stem cell derived CMs maintain a relative high proliferation potential and expression of proliferation-associated markers such a Ki-67 [40]. This indicates that the "CMs outgrowth phenotype" monitored in our assay is likely attributed to both cardiomyocyte proliferation as well as motility. Although our assay covers the result of both such effects and was mainly developed as a primary screen to identify compounds and respective concentrations, a detailed analysis to distinguish between these phenomena is necessary to reveal molecular mechanism(s) of specific compounds. Thus, our focus here is on developing a novel assay strategy for measuring drug effects which are not covered by other applied assays and which provide valuable information on novel, potentially unexpected effects of respective drugs. In that way, the outgrowth assay endpoints have a clear relevance for drug assessment.

Importantly, the data in our study were closely recapitulated by using CMs derived from two independent hiPSC lines, strongly supporting robustness and general (rather than cell line dependent) relevance of the method and its results.

Finally, aggregates of cardiomyocytes cultured under dynamic conditions have been reported to grow in size, contain more nuclei, and display increased contraction forces compared to static cultures. However, we did not observe any (potentially expected) differences between static and dynamic culture conditions in our assay, although a clearly observable gravity-induced medium flow was achieved via the microfluidic design on the rocker-device. Further studies for dynamic studies may implement a higher shear rate/stress compared to our approach (mean value approximately 4.5 dyn/cm^2). This can be achieved, for example, by connecting an external pump or by increasing the height of the well to create a higher velocity of the fluid through the channel. This may be a critical issue if the size of the microchannels are smaller than in those in this report. On the other hand, it remains open how such experimental flow conditions will compare e.g., to tissue perfusion in an organ such as the heart.

5. Conclusions

In this paper, the number of cells surrounding cardiac spheroids attached to the surfaces of perfused microfluidic channels was used as a quantitative measurement of the effect of different compounds. By staining the cell nuclei with a fluorescent marker, it was possible to induce both a decrease (doxorubicin and endothelin-1) and an increase (phenylephrine and amiodarone) in number of cells surrounding the spheroids.

Based on these findings, we suggest that this measurement procedure is useful as an assay for evaluating drug molecules in a 3D environment. However, the assay should preferably be performed as a complement to already existing end-point assays for providing supportive data. This could for example include beating rate.

The assay is carried out in a spheroid-on-a-chip device. A few caveats should be mentioned with this format. Drug molecules of varying hydrophobicity may be affected by surface properties of the fabrication materials of device. Also, the conditions for the cells grown out from the spheroids in the microfluidic channel may not be fully representative for the 3D model. These could be parameters to be further evaluated in validation of the assay.

The spheroid chip format is cost-effective, requires short training before practice, and can easily be scaled-up to higher throughout for routine laboratory work. This makes the spheroid-on-a-chip a convenient complementary assay tool for drug development.

Supplementary Materials: The following are available online at http://www.mdpi.com/2306-5354/5/2/36/s1, Figure S1: Schematics showing the dynamic condition of the spheroid-on-a-chip device on the rocker setup. The frequency of the rocker was 0.5 Hertz, or 30 alternating cycles per minute, Figure S2: Depiction of the different areas detected by the high content imaging system. The blue area represents the attached spheroid. The red circle shows the area surrounding the spheroid which was excluded from the analysis to avoid any counting of cells which were still located in the spheroid. The green area represents the analyzed area for analysis. Hoechst 33342 positive nuclei in this area were counted and normalized to the area of the green circle.

Author Contributions: Conceptualization, J.C. and F.M.; Experiments, J.C., F.M., H.K.; K.S. and M.C.; Resources, M.B.; Writing, J.C. C.-F.M., R.Z. and F.M.; Writing-review & editing, M.B.

Acknowledgments: The research leading to these results has received support from the Innovative Medicines Initiative Joint Undertaking under grant agreement no. 115439 to the project StemBANCC, resources of which are composed of financial contribution from the European Union's Seventh Framework Programme (FP7/2007-2013) and EFPIA companies' in kind contribution. The project was also funded by the Swedish Research Council (Vetenskapsrådet, grant 2015-03298). Additional project related funding to RZ includes support by the German Research Foundation (DFG; Cluster of Excellence REBIRTH EXC 62/2, ZW64/4-1 as well as KFO311 and

ZW64/7-1), by the German Ministry for Education and Science (BMBF, grants: 13N14086, 01EK1601A, 01EK1602A), and the European Union H2020 program to the project TECHNOBEAT (grant 66724). This publication reflects only the author's views and neither the IMI JU nor EFPIA nor the European Commission are liable for any use that may be made of the information contained therein.

Conflicts of Interest: The authors declare no conflict of interest.

References

1. Bhatia, S.N.; Ingber, D.E. Microfluidic organs-on-chips. *Nat. Biotechnol.* **2014**, *32*, 760–772. [CrossRef] [PubMed]

2. Sung, J.H.; Esch, M.B.; Prot, J.-M.; Long, C.J.; Smith, A.; Hickman, J.J.; Shuler, M.L. Microfabricated mammalian organ systems and their integration into models of whole animals and humans. *Lab Chip* **2013**, *13*, 1201–1212. [CrossRef] [PubMed]

3. Prodanov, L.; Jindal, R.; Bale, S.; Hegde, M.; McCarty, W.J.; Golberg, I.; Bhushan, A.; Yarmush, M.L.; Usta, O.B. Long-term maintenance of a microfluidic 3D human liver sinusoid. *Biotechnol. Bioeng.* **2016**, *113*, 241–246. [CrossRef] [PubMed]

4. Lanz, H.L.; Saleh, A.; Kramer, B.; Cains, J.; Ng, C.P.; Trietsch, S.J.; Hankemeier, T.; Joore, J.; Vulto, P.; Weinshilboum, R.; et al. Therapy response testing of breast cancer in a 3D high-throughput perfused microfluidic platform. *BMC Cancer* **2017**, *17*, 709. [CrossRef] [PubMed]

5. Ulrich, E.; Patch, C.; Aigner, S.; Graf, M.; Iacone, R.; Freskgård, P.O. Multicellular self-assembled spheroidal model of the blood brain barrier. *Sci. Rep.* **2013**, *3*, 1500. [CrossRef] [PubMed]

6. Bell, C.C.; Lauschke, V.M.; Vorrink, S.U.; Palmgren, H.; Duffin, R.; Andersson, T.B.; Ingelman-Sundberg, M. Transcriptional, functional, and mechanistic comparisons of stem cell-derived hepatocytes, HepaRG cells, and three-dimensional human hepatocyte spheroids as predictive in vitro systems for drug-induced liver injury. *Drug Metab. Dispos.* **2017**, *45*, 419–429. [CrossRef] [PubMed]

7. Mueller, D.; Krämer, L.; Hoffmann, E.; Klein, S.; Noor, F. 3D organotypic HepaRG cultures as in vitro model for acute and repeated dose toxicity studies. *Toxicol. In Vitro* **2014**, *28*, 104–112. [CrossRef] [PubMed]

8. Andersson, H.; Steel, D.; Asp, J.; Dahlenborg, K.; Jonsson, M.; Jeppsson, A.; Lindahl, A.; Kågedal, B.; Sartipy, P.; Mandenius, C.-F. Assaying cardiac biomarkers for toxicity testing using biosensing and cardiomyocytes derived from human embryonic stem cells. *J. Biotechnol.* **2010**, *150*, 175–181. [CrossRef] [PubMed]

9. Bistola, V.; Nikolopoulou, M.; Derventzi, A.; Kataki, A.; Sfyras, N.; Nikou, N.; Toutouza, M.; Toutouzas, P.; Stefanadis, C.; Konstadoulakis, M.M. Long-term primary cultures of human adult atrial cardiac myocytes: Cell viability, structural properties and bnp secretion in vitro. *Int. J. Cardiol.* **2008**, *131*, 113–122. [CrossRef] [PubMed]

10. Natarajan, A.; Stancescu, M.; Dhir, V.; Armstrong, C.; Sommerhage, F.; Hickman, J.J.; Molnar, P. Patterned cardiomyocytes on microelectrode arrays as a functional, high information content drug screening platform. *Biomaterials* **2011**, *32*, 4267–4274. [CrossRef] [PubMed]

11. Bergström, G.; Christoffersson, J.; Schwanke, K.; Zweigerdt, R.; Mandenius, C.-F. Stem cell derived in vivo-like human cardiac bodies in a microfluidic device for toxicity testing by beating frequency imaging. *Lab Chip* **2015**, *15*, 3242–3249. [CrossRef] [PubMed]

12. Frimat, J.-P.; Sisnaiske, J.; Subbiah, S.; Menne, H.; Godoy, P.; Lampen, P.; Leist, M.; Franzke, J.; Hengstler, J.G.; van Thriel, C.; et al. The network formation assay: A spatially standardized neurite outgrowth analytical display for neurotoxicity screening. *Lab Chip* **2010**, *10*, 701–709. [CrossRef] [PubMed]

13. Van der Meer, A.D.; Vermeul, K.; Poot, A.A.; Feijen, J.; Vermes, I. A microfluidic wound-healing assay for quantifying endothelial cell migration. *Am. J. Physiol. Heart Circ. Physiol.* **2009**, *298*, H719–H725. [CrossRef] [PubMed]

14. Zeng, W.R.; Beh, S.-J.; Bryson-Richardson, R.J.; Doran, P.M. Production of zebrafish cardiospheres and cardiac progenitor cells in vitro and three-dimensional culture of adult zebrafish cardiac tissue in scaffolds. *Biotechnol. Bioeng.* **2017**, *114*, 2142–2148. [CrossRef] [PubMed]

15. Jang, K.-J.; Mehr, A.P.; Hamilton, G.A.; McPartlin, L.A.; Chung, S.; Suh, K.-Y.; Ingber, D.E. Human kidney proximal tubule-on-a-chip for drug transport and nephrotoxicity assessment. *Integr. Biol.* **2013**, *5*, 1119–1129. [CrossRef] [PubMed]

16. Booth, R.; Kim, H. Characterization of a microfluidic in vitro model of the blood-brain barrier (μBBB). *Lab Chip* **2012**, *12*, 1784–1792. [CrossRef] [PubMed]

17. Esch, M.B.; Prot, J.M.; Wang, Y.I.; Miller, P.; Llamas-Vidales, J.R.; Naughton, B.A.; Applegate, D.R.; Shuler, M.L. Multi-cellular 3D human primary liver cell culture elevates metabolic activity under fluidic flow. *Lab Chip* **2015**, *15*, 2269–2277. [CrossRef] [PubMed]

18. Jackman, C.P.; Carlson, A.L.; Bursac, N. Dynamic culture yields engineered myocardium with near-adult functional output. *Biomaterials* **2016**, *111*, 66–79. [CrossRef] [PubMed]

19. Morrison, M.; Klein, C.; Clemann, N.; Collier, D.A.; Hardy, J.; Heißerer, B.; Cader, M.Z.; Graf, M.; Kaye, J. Stembancc: Governing access to material and data in a large stem cell research consortium. *Stem Cell Rev. Rep.* **2015**, *11*, 681–687. [CrossRef] [PubMed]

20. Kempf, H.; Olmer, R.; Kropp, C.; Ruckert, M.; Jara-Avaca, M.; Robles-Diaz, D.; Franke, A.; Elliott, D.A.; Wojciechowski, D.; Fischer, M.; et al. Controlling expansion and cardiomyogenic differentiation of human pluripotent stem cells in scalable suspension culture. *Stem Cell Rep.* **2014**, *3*, 1132–1146. [CrossRef] [PubMed]

21. Kempf, H.; Kropp, C.; Olmer, R.; Martin, U.; Zweigerdt, R. Cardiac differentiation of human pluripotent stem cells in scalable suspension culture. *Nat. Protoc.* **2015**, *10*, 1345–1361. [CrossRef] [PubMed]

22. Kempf, H.; Olmer, R.; Haase, A.; Franke, A.; Bolesani, E.; Schwanke, K.; Robles-Diaz, D.; Coffee, M.; Göhring, G.; Dräger, G.; et al. Bulk cell density and Wnt/TGFbeta signalling regulate mesendodermal patterning of human pluripotent stem cells. *Nat. Commun.* **2016**, *7*, 13602. [CrossRef] [PubMed]

23. Maillet, A.; Tan, K.; Chai, X.; Sadananda, S.N.; Mehta, A.; Ooi, J.; Hayden, M.R.; Pouladi, M.A.; Ghosh, S.; Shim, W.; et al. Modeling doxorubicin-induced cardiotoxicity in human pluripotent stem cell derived-cardiomyocytes. *Sci. Rep.* **2016**, *6*, 25333. [CrossRef] [PubMed]

24. Carlson, C.; Koonce, C.; Aoyama, N.; Einhorn, S.; Fiene, S.; Arne, T.; Swanson, B.; Anson, B.; Kattman, S. Phenotypic Screening with Human iPS Cell–Derived Cardiomyocytes: HTS-Compatible Assays for Interrogating Cardiac Hypertrophy. *J. Biomol. Screen.* **2013**, *18*, 1203–1211. [CrossRef] [PubMed]

25. Vane, J.R.; Botting, R.M. The mechanism of action of aspirin. *Thromb. Res.* **2003**, *110*, 255–258. [CrossRef]

26. Schäfer, M.; Frischkopf, K.; Taimor, G.; Piper, H.M.; Schlüter, K.-D. Hypertrophic effect of selective $β_1$-adrenoceptor stimulation on ventricular cardiomyocytes from adult rat. *Am. J. Physiol.-Cell Physiol.* **2000**, *279*, C495–C503. [CrossRef] [PubMed]

27. Ogoh, S.; Sato, K.; Fisher, J.P.; Seifert, T.; Overgaard, M.; Secher, N.H. The effect of phenylephrine on arterial and venous cerebral blood flow in healthy subjects. *Clin. Physiol. Funct. Imaging* **2011**, *31*, 445–451. [CrossRef] [PubMed]

28. Lalevée, N.; Barrère-lemaire, S.; Gautier, P.; Nargeot, J.; Richard, S. Effects of amiodarone and dronedarone on voltage-dependent sodium current in human cardiomyocytes. *J. Cardiovasc. Electrophysiol.* **2003**, *14*, 885–890. [CrossRef] [PubMed]

29. Tanaka, A.; Yuasa, S.; Mearini, G.; Egashira, T.; Seki, T.; Kodaira, M.; Kusumoto, D.; Kuroda, Y.; Okata, S.; Suzuki, T.; et al. Endothelin-1 induces myofibrillar disarray and contractile vector variability in hypertrophic cardiomyopathy–induced pluripotent stem cell–derived cardiomyocytes. *J. Am. Heart Assoc.* **2014**, *3*, e001263. [CrossRef] [PubMed]

30. Suzuki, T.; Miyauchi, T. A novel pharmacological action of ET-1 to prevent the cytotoxicity of doxorubicin in cardiomyocytes. *Am. J. Physiol.-Regul. Integr. Comp. Physiol.* **2001**, *280*, R1399–R1406. [CrossRef] [PubMed]

31. Taylor, J.M.; Rovin, J.D.; Parsons, J.T. A role for focal adhesion kinase in phenylephrine-induced hypertrophy of rat ventricular cardiomyocytes. *J. Biol. Chem.* **2000**, *275*, 19250–19257. [CrossRef] [PubMed]

32. Dempsey, G.T.; Chaudhary, K.W.; Atwater, N.; Nguyen, C.; Brown, B.S.; McNeish, J.D.; Cohen, A.E.; Kralj, J.M. Cardiotoxicity screening with simultaneous optogenetic pacing, voltage imaging and calcium imaging. *J. Pharmacol. Toxicol. Methods* **2016**, *81*, 240–250. [CrossRef] [PubMed]

33. Yokoo, N.; Baba, S.; Kaichi, S.; Niwa, A.; Mima, T.; Doi, H.; Yamanaka, S.; Nakahata, T.; Heike, T. The effects of cardioactive drugs on cardiomyocytes derived from human induced pluripotent stem cells. *Biochem. Biophys. Res. Commun.* **2009**, *387*, 482–488. [CrossRef] [PubMed]

34. Mehta, A.; Chung, Y.; Sequiera, G.L.; Wong, P.; Liew, R.; Shim, W. Pharmacoelectrophysiology of viral-free induced pluripotent stem cell–derived human cardiomyocytes. *Toxicol. Sci.* **2013**, *131*, 458–469. [CrossRef] [PubMed]

35. Zweigerdt, R. The art of cobbling a running pump—Will human embryonic stem cells mend broken hearts? *Semin. Cell Dev. Biol.* **2007**, *18*, 794–804. [CrossRef] [PubMed]

36. Laflamme, M.A.; Chen, K.Y.; Naumova, A.V.; Muskheli, V.; Fugate, J.A.; Dupras, S.K.; Reinecke, H.; Xu, C.; Hassanipour, M.; Police, S.; et al. Cardiomyocytes derived from human embryonic stem cells in pro-survival factors enhance function of infarcted rat hearts. *Nat. Biotechnol.* **2007**, *25*, 1015. [CrossRef] [PubMed]

37. Rojas, S.V.; Martens, A.; Zweigerdt, R.; Baraki, H.; Rathert, C.; Schecker, N.; Rojas-Hernandez, S.; Schwanke, K.; Martin, U.; Haverich, A.; et al. Transplantation effectiveness of induced pluripotent stem cells is improved by a fibrinogen biomatrix in an experimental model of ischemic heart failure. *Tissue Eng. Part A* **2015**, *21*, 1991–2000. [CrossRef] [PubMed]

38. Van den Akker, F.; Feyen, D.A.M.; van den Hoogen, P.; van Laake, L.W.; van Eeuwijk, E.C.M.; Hoefer, I.; Pasterkamp, G.; Chamuleau, S.A.J.; Grundeman, P.F.; Doevendans, P.A.; et al. Intramyocardial stem cell injection: Go(ne) with the flow. *Eur. Heart J.* **2017**, *38*, 184–186. [CrossRef] [PubMed]

39. Lian, X.; Hsiao, C.; Wilson, G.; Zhu, K.; Hazeltine, L.B.; Azarin, S.M.; Raval, K.K.; Zhang, J.; Kamp, T.J.; Palecek, S.P. Robust cardiomyocyte differentiation from human pluripotent stem cells via temporal modulation of canonical Wnt signaling. *Proc. Natl. Acad. Sci. USA* **2012**, *109*, E1848–E1857. [CrossRef] [PubMed]

40. Weber, N.; Schwanke, K.; Greten, S.; Wendland, M.; Iorga, B.; Fischer, M.; Geers-Knörr, C.; Hegermann, J.; Wrede, C.; Fiedler, J.; et al. Stiff matrix induces switch to pure β-cardiac myosin heavy chain expression in human ESC-derived cardiomyocytes. *Basic Res. Cardiol.* **2016**, *111*, 68. [CrossRef] [PubMed]

41. Iorga, B.; Schwanke, K.; Weber, N.; Wendland, M.; Greten, S.; Piep, B.; dos Remedios, C.G.; Martin, U.; Zweigerdt, R.; Kraft, T.; et al. Differences in contractile function of myofibrils within human embryonic stem cell-derived cardiomyocytes vs. adult ventricular myofibrils are related to distinct sarcomeric protein isoforms. *Front. Physiol.* **2018**, *8*, 1111. [CrossRef] [PubMed]

bioengineering

MDPI

Article

Efficient Computational Design of a Scaffold for Cartilage Cell Regeneration

Tannaz Tajsoleiman [1], Mohammad Jafar Abdekhodaie [2], Krist V. Gernaey [1] and Ulrich Krühne [1,*]

[1] Department of Chemical and Biochemical Engineering, Technical University of Denmark, DK-2800 Kgs., Lyngby, Denmark; tantaj@kt.dtu.dk (T.T.); kvg@kt.dtu.dk (K.V.G.)
[2] Department of Chemical and Petroleum Engineering, Sharif University of Technology, Tehran, Iran; abdmj@sharif.edu
[*] Correspondence: ulkr@kt.dtu.dk; Tel.: +45-4525-2960

Received: 8 March 2018; Accepted: 20 April 2018; Published: 24 April 2018

Abstract: Due to the sensitivity of mammalian cell cultures, understanding the influence of operating conditions during a tissue generation procedure is crucial. In this regard, a detailed study of scaffold based cell culture under a perfusion flow is presented with the aid of mathematical modelling and computational fluid dynamics (CFD). With respect to the complexity of the case study, this work focuses solely on the effect of nutrient and metabolite concentrations, and the possible influence of fluid-induced shear stress on a targeted cell (cartilage) culture. The simulation set up gives the possibility of predicting the cell culture behavior under various operating conditions and scaffold designs. Thereby, the exploitation of the predictive simulation into a newly developed stochastic routine provides the opportunity of exploring improved scaffold geometry designs. This approach was applied on a common type of fibrous structure in order to increase the process efficiencies compared with the regular used formats. The suggested topology supplies a larger effective surface for cell attachment compared to the reference design while the level of shear stress is kept at the positive range of effect. Moreover, significant improvement of mass transfer is predicted for the suggested topology.

Keywords: tissue engineering; CFD simulation; scaffold geometry optimization; micro-bioreactor operating conditions

1. Introduction

Cartilage diseases such as hyaline damages are among the most common skeletal health issues [1,2]. In case of advanced cartilage diseases, there is a considerable need for an external intervention due to the limited recovery ability of mature cartilage cells (chondrocyte), particularly for elderly patients. In an advanced case of tissue damage, when partial or complete cartilage transplantation is needed, tissue engineering demonstrates its potential as a candidate alternative for auto-graft/allograft tissue transplantation. In a successful tissue engineering procedure, physical and physiological properties of the target tissue are recreated into a new 3 dimensional cell structure during a complex cell culture process [3].

Optimized protocols and operating conditions can guarantee a successful tissue regeneration cell culture. The term operating condition is a general description, which covers the capacity of mass transfer as a function of cell culture strategy and physical environment characteristics influenced by the type of reactor. Mammalian cell cultures are generally operated under either static or dynamic cultivation principles. Mass transfer capacity of a static cell culture is mainly controlled by molecular diffusion within the growth environment [4]. Lack of nutrient supply and an increased chance of metabolite accumulation are the main bottlenecks in high-density static cell cultures. Using perfusion bioreactors is an alternative approach to increase mass transfer within a cell culture environment.

In the case of mammalian cell cultures, shear stress can boost cell metabolism and proliferation rate [5–8]. This parameter is directly controlled by the induced flows in perfusion bioreactors. All these considerations illustrate the complexity of a dynamic cell culture and the necessity of having a comprehensive knowledge of the hydrodynamic environment of the cell culture.

Present limitations of on-line monitoring techniques, particularly at the cell level, form one of the most persuading reasons to use mathematical modelling and simulation to estimate process variables. For instance, Williams et al. used computational fluid dynamics (CFD) simulation to quantify momentum and mass transfer within a cartilage cell culture in order to improve the design of a concentric cylinder-shape bioreactor [9]. In their work, they focused on flow field, shear stress profile and molecular oxygen distribution around a non-porous structure, represented as a three-dimensional (3D) cell growth space. Raimondi et al. used computational modelling to estimate fluid dynamic properties induced on a mature cartilage cell to characterize the effect of perfusion flow on the cell culture [10]. The coupled experimental model was extracted based on a dynamic cell culture on a biodegradable scaffold, which was exposed to a bi-directional flow of culture medium. Krühne et al. tried to directly use CFD to simulate the dynamic interactions between the liquid flow and growing cells in an exemplary pore, as a part of a porous scaffold. In this regard, they introduced a simple biological growth model based on Michaelis-Menten kinetics, combined with a shear stress term, to estimate the cell growth rate under fluid flow conditions [11]. Later, mathematical modelling and computational simulation were extensively used to study environmental conditions in various cell cultures [12,13]. Mathematical modelling and CFD simulations are not only used for characterizing a process behavior, but they are also helpful tools to optimize a process condition and improve the culture environmental design. This study presents a systematic approach to use computational predictions for more efficient 3D cell cultures. Hence, the first part of this paper focuses on mathematical modelling and CFD simulation of a 3D scaffold based cartilage cell culture within a perfusion bioreactor. In this part, a predictive model is introduced by considering various biochemical and physical factors. This model covers the main effective parameters such as nutrient supply, metabolite concentration and fluid shear stress on cartilage cells within a wide range of environmental conditions. The considered perfusion reactor and the investigated scaffold are shown in Figure 1. The developed model potentially helps to understand and characterize the cell culture behavior under perfusion flow conditions.

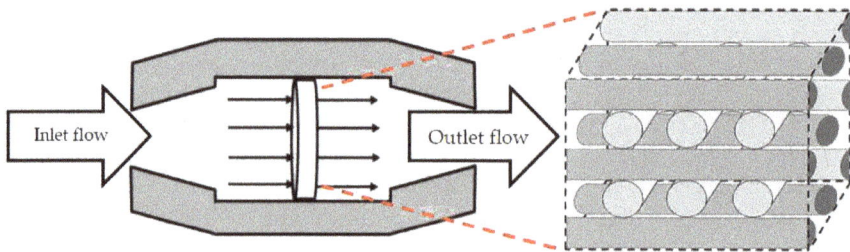

Figure 1. 3D fibrous scaffold placed in a perfusion micro-bioreactor.

The studied scaffold is a common type of fibrous structure that is frequently reported for 3D cartilage regeneration [12,14–16]. This structure can be fabricated by assembling polymeric fibers with a repeated pattern, or by using 3D printers. Hossain et al. initially discussed the fluid velocity drop, downstream of the rear side of the fibers, as a design weakness of this structure. This fact is a considerable phenomenon, particularly in the case of relatively slow perfusion or low nutrient concentrations in the feed stream [17]. Hence, the effect of the scaffold design on the cell culture process needs to be investigated in more detail. Here, the scaffold design is represented by the fiber diameters and the compaction depth h_c at the attachment point between two fibers, which is formulated into the attachment angle α in Figure 2. The second part of the article focuses on using a CFD model to reach an improved scaffold design. Thereby, a new stochastic topology optimization algorithm is introduced to build a bridge between the characterization step and environmental design corrections and improvements. Accordingly, improved design dimensions for this class of scaffolds are proposed with respect to the fiber diameters and the attachment angles. In the last part of the paper, the effect of bioreactor operating conditions, such as perfusion velocity and medium nutrient concentration, on the culture efficiency is studied.

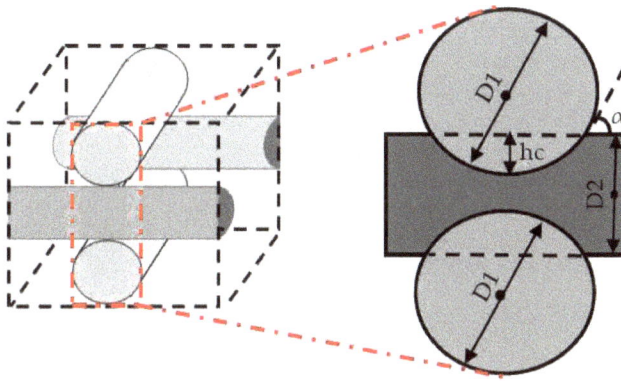

Figure 2. Scaffold design characteristics at the attachment point.

2. Mathematical Model

In order to predict a cartilage cell culture, a multi-phase model is developed by including three major domains, i.e., the solid fibrous scaffold, the perfusion medium and the dynamic biomass volume. The scaffold is assumed as a rigid impermeable structure, which provides cell attachment surfaces. This assumption simplifies the predictive model by eliminating mass transfer phenomena inside the solid phase. Furthermore, simulating the medium properties with the assumption of Newtonian behavior significantly reduces the complexity of the equations. In this model, the biomass domain represents a dynamic heterogeneous volume, attached to the solid surface, with various ratios of medium and cells volume fractions (ε_f and ε_c, respectively). By changing the cell density during the cultivation, the ratio of volume fractions within the biomass domain changes as a function of time and location according to Equation (1). This equation is formulated assuming a constant solid phase volume.

$$\varepsilon_f (x, y, z, t) + \varepsilon_c (x, y, z, t) = 1, \tag{1}$$

The distributions of nutrients and metabolite concentrations in the biomass domain are mainly controlled by ε_c and ε_f. The biochemical aspects of various nutrient concentrations in cartilage cell cultures were extensively investigated in previous studies [8,18,19]. For instance, oxygen and glucose are identified as the main nutrient supplies for mammalian cell metabolism. It is established that the oxygen uptake rate level in cartilage cell cultures is adjusted based on the glucose consumption rate [3,20,21]. Hence, glucose was selected as the main nutrient to be included into the predictive model. The distribution of glucose concentrations is calculated according to Equation (2).

$$\frac{\partial}{\partial t} \left[\varepsilon_f \times c_g^f + \varepsilon_c \times c_g^c \right] = \nabla \times \left(D_g^f \times \nabla c_g^f + D_g^c \times \nabla c_g^c \right) - R_g - v \nabla \times c_g^f, \tag{2}$$

Given Equation (2), c_g^f and c_g^c are the intrinsic average glucose concentrations, and D_g^f and D_g^c are diffusion tensors in the fluid (f) and the cell (c) respectively. It should be mentioned that intercellular mass transport is restricted to molecular diffusion as a function of cell physiology. The glucose consumption rate R_g is formulated according to Michaelis-Menten (M-M) kinetics in Equation (3), both as a function of cell density and intercellular glucose concentration:

$$R_g = \frac{V_m \, \varepsilon_b \times c_g{}^c}{K_m + \varepsilon_b \times c_g{}^c},$$

(3)

where, V_m and K_m are the maximum specific glucose uptake rate and the half-saturation constant for glucose, respectively. The interfacial properties for the glucose concentration at the cell membrane were simplified by introducing the linear equilibrium coefficient K_{eq} [22].

In this model, glucose breakdown resulting into lactate formation is considered as the main metabolic pathway for cartilage cells [20,21]. Governing equation of lactate concentration in fluid phase $c_l{}^f$ is shown in Equation (4).

$$\frac{\partial}{\partial t}\left[c_l{}^f\right] = \nabla \times \left(D_l{}^f \times \nabla c_l{}^f\right) + R_l - v\,\nabla \times c_l{}^f,$$

(4)

Rate of lactate formation R_l is estimated to be two times of glucose consumption rate. The value of the lactate concentration is directly reflected by the pH level of the cell culture according to Equation (5) [23,24].

$$\text{pH} = 7.4 - 0.0406\, C_l,$$

(5)

It should be considered that glucose has a double effect on cartilage cell culture. Lack of glucose essentially threatens the cell viability and growth rate, whereas an increase in glucose concentration proportionally boosts the cell metabolism rate and the correlated lactate production rate. Toxicity is the main outcome of a high metabolite concentration that is observed by dropping pH levels in an inefficient cell culture environment. Various cell types show different pH resistance [25] Cartilage cells are sensitive to an acidic environment with a pH value lower than 6.8. This condition is formulated as a step function S{pH} and the value of ε_c is updated to $\varepsilon_c \times S\{pH\}$.

Mathematical modelling also provides the possibility of estimating the cell number and distribution. The cultured cell population is mainly controlled by three terms: proliferation, death and migration. The dynamic cell distribution during a cell culture can be quantified into ε_b according to Equation (6) [26].

$$\frac{\partial}{\partial t}\left[\varepsilon_c \times \rho_c{}^b\right] = \nabla \times \left(D_c \times \nabla\left(\varepsilon_c \times \rho_c{}^b\right)\right) + \left(r_g - r_d\right) \times \varepsilon_c \times \rho_c{}^b,$$

(6)

where, r_g and r_d are cell growth and death rate, respectively. $\rho_c{}^b$ is the intrinsic average cell mass density and D_c is the cell effective diffusion tensor that indicates the migration term. The migration term is quantitatively negligible compared with the two other terms. Nutrients concentration, metabolite distribution and physical microenvironment properties (e.g., pH and fluid shear stress) are the main controlling parameters of the cell proliferation rate. In this study, the effect of metabolite concentration on cell culture is considered solely from an environmental (pH limiting range) point of view. The effect of the glucose concentration, cell density and local shear stress on the proliferation rate is given by Equation (7) [27].

$$r_g = \frac{K_g \times c_g{}^c}{K_c \times \rho_{cell}\varepsilon_c + c_g{}^c},$$

(7)

where, K_c and ρ_{cell} are the saturation coefficient and cellular density respectively. In this equation, K_g represents the effect of fluid shear stress on the cell proliferation rate. The stimulating influence of low range shear stress on the cell growth rate has been established in previous studies [3,12,13,27,28]. However, this effect changes to an inhibitory factor at a higher range of shear stress values, such that shear stress higher than 1 Pa has been reported as a damaging condition for the cell viability [12,13].

Equation (8) shows a modified correlation between the level of fluid induced shear stress and the cartilage cell growth rate.

$$K_g = \begin{cases} K_{g_0}(\alpha + \beta\tau) & for\ \tau \in [0,0.1)\,Pa \\ K_{g_0} \times 11.326 & for\ \tau \in [0.1,0.6)\,Pa \\ K_{g_0} \times (2.5 \times (1-\tau) \times 11.326) & for\ \tau \in [0.6,1)\,Pa \\ 0 & for\ \tau > 1\,Pa \end{cases} \quad (8)$$

In Equation (10), K_{g_0} is the maximum growth rate under static culture conditions and α and β are constant parameters which are illustrated in Table 1.

A transient scaffold based cartilage cell culture was simulated based on the above-mentioned model coupled with the Navier-Stokes and continuity equations [29]. The hydrodynamic behavior of the perfusion flow was predicted by applying a laminar model on a representative Newtonian fluid (water properties were considered for the description of the fluid phase). The boundary conditions for the fluid phase were set to a plug flow at the inlet interface, an opening outlet with relatively zero pressure and no-slip condition at the scaffold walls. The simulation was initialized by specifying the biomass domain as a uniform volume, with the viscosity of μ_c and the thickness of d_c over the solid attachment surface. The local value of μ_c was set as a linear function of the cell volume fraction. By growing the cells population, the local value of μ_c increases proportionally with ε_c within the range of 0.001 Pa.s, as the presence of no cells, to 1 Pa.s as a fully packed bulk of the cells. The calculated viscosity regulates the penetration rate of the fluid in the biomass domain. Moreover, the assumption of a monolayer cell culture was included by setting the value of d_c to the dimension of one cartilage cell and limiting the upper bond of the cells distribution (cell/attachment area), according to the available attachment surface. The CFD simulation was performed using second order backward Euler approach and finally, the results were validated by available experimental data in the literature [30].

Table 1. Numerical values of the model parameters used in the simulations.

Definition	Value	Reference
Cell density, ρ_{cell}	0.182 [g/cm^3]	[13]
Glucose diffusion tensor in fluid, $D_g{}^f$	1×10^{-5} [cm^2/s]	[31]
Glucose diffusion tensor in the cell, $D_g{}^c$	1×10^{-6} [cm^2/s]	[31]
Glucose diffusion tensor in the cell, $D_f{}^f$	1.4×10^{-5} [cm^2/s]	[23]
Glucose maximal consumption rate, V_m	3.9×10^{-5} [kg/m^{3}s]	[32]
Glucose half saturation constant, K_m	6.3×10^{-3} [kg/m^3]	[32]
K_{g_0}	5.8×10^{-6} [1/s]	[13]
Equilibrium coefficient, K_{eq}	0.1	[26]
α	0.8761	[13]
β	0.1045×10^3 [1/Pa]	[13]
d_c	13 [µm]	[33]

3. Methodology

As discussed in the introduction, the design of a scaffold and bioreactor operating conditions are critical parameters, which have to guarantee a balance between nutrient supply rate, metabolite removal performance and physical environmental conditions. Modelling and simulation are powerful tools for estimating the level of effective factors—such as nutrients concentrations and other environmental factors like shear stress—to predict the state of a cell culture under various operating conditions. One of the main advantages of process simulation is that it offers the opportunity to better understand the ongoing phenomena and diagnose the design weaknesses and the potential bottlenecks towards achieving an optimal design. In this overall perspective, the geometrical aspects of the scaffold design are investigated in order to systematically explore an improved design for more efficient cartilage cell cultures. Hence, a CFD based stochastic algorithm was developed according to the following steps:

Step 1: Division of the studied geometry into a network of repeated sections towards focusing only on a single unit due to the intensive computational demands. The selected unit has to include the main influence on the fluid dynamic conditions. Thus, the studied geometry is initially discretized into its structural repetition element (Figure 3b). Figure 3c shows a section of the fluid volume where the main hydrodynamic fluid phenomena are happening. This volume is labelled as the investigation 'unit' that is considered for detailed geometrical studies.

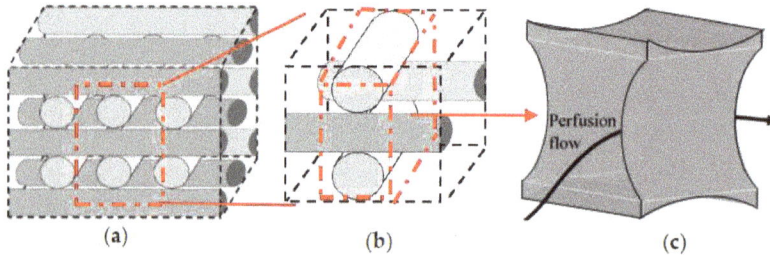

Figure 3. Geometry simplification (**a**) full structure, (**b**) conjunction of four fibers, (**c**) the studied simulation volume (unit).

Step 2: Identification of an indicator parameter to quantitatively evaluate the overall cell culture conditions. In this case study, the dimensionless parameter 'culture efficiency (CE)' is introduced to evaluate the yield of the cell culture under various geometrical and operating conditions. The culture efficiency is defined in Equation (9).

$$CE = \frac{\overline{(cell\ density)}_{final} - \overline{(cell\ density)}_{inital}}{\overline{(cell\ density)}_{initial}},\qquad(9)$$

Step 3: Evaluation of various design possibilities with respect to the indicator parameter and fabrication feasibilities. The final state of the cell culture is mainly controlled by the mass transfer capacity into the culture environment and the level of hydrodynamically induced shear stress on the cells. On the other hand, for monolayer cell cultures such as for cartilage cells, the available amount of attachment surfaces for the growing biomass is a deterministic parameter [34,35]. Hence, the main challenge is to find an improved scaffold design which provides an efficient mass transfer over an extended attachment surface, while simultaneously keeping the level of shear stress at the supportive range of effect.

The estimation of the indicator parameter (CE) through the CFD simulation gives the opportunity of assessing the capabilities of various structures to fulfil a desired cultivation condition during a cell culture. Regarding the intensive computational efforts experienced when considering the complexity of the process for each scaffold candidate, the evaluation of all possible structures is not feasible. Therefore, a finite number of candidate structures are selected through a new stochastic method to reach an improved structure. Each candidate structure is specified by initially introducing an imaginary peripheral cube with dimension $2r_1$, which assigns the maximum possible volume for one unit. This cube is fixed by the allocation of 8 points at its corners, as shown in Figure 4. In this unit, the front and back fibers (fiber 3 and fiber 4) are named as the secondary fibers, which are positioned in relation to the top and bottom fiber diameters (fiber 1 and fiber 2; primary fibers) and the compaction depth. The diameters of the primary fibers are defined by introducing indicator points 1 and 2 at the center line of the cube's front face in the Y axis direction. The normal distance Δ between each indicator parameter and the center of the closest imaginary cube's edge illustrate the corresponding fiber's diameter D according to Equation (10). Accordingly, secondary fibers are specified by indicating the coordinates of points 3 and 4 with respect to the closest imaginary edge at the cube's left face. Movement of the indicator points from the cube's edge in their corresponding faces (in the Y direction for points 1 and 2, and in the Z direction for points 3 and 4) indicates the main design characteristic parameters of various possible structures. The movements are specified by setting the displacement values Δ s. Moreover, 12 extra points, named 'free point', are used to illustrate the rest of the corners and edges in the simulation software. The locations of the free points are adjusted according to the coordinates of the indicator parameters in the three-dimensional space.

$$D = \frac{r^2 + \Delta^2}{\Delta}, \tag{10}$$

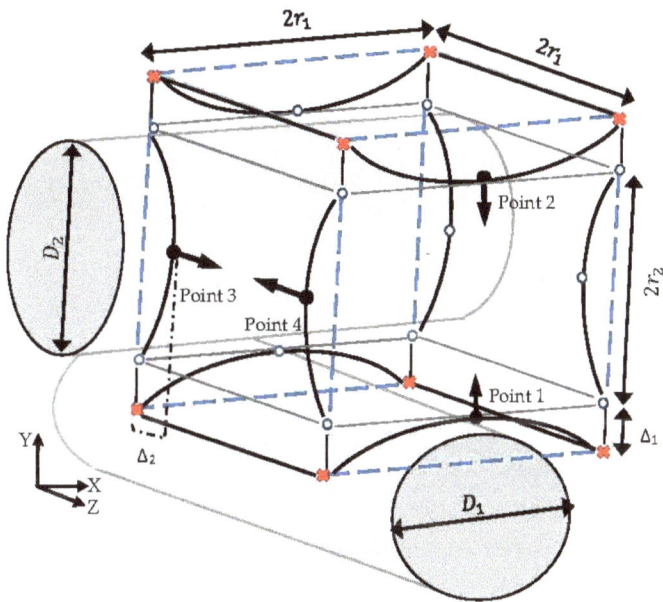

Figure 4. The studied unit: ✖ fixed point, ● indicator point, ○ free point, – - imaginary cube edges.

To reach a feasible design, considering fabrication issues, it is assumed that the indicator points at the same face take similar displacement values Δ. The displacement value Δ is constrained by the maximum possible volume of one unit (volume of the imaginary cube). In this case study, the dimension of the imaginary cube was assumed to be 200 μm. Moreover, a random array of uniform distributed Δ was generated in the range of 0–100 μm, in order to specify and study a limited number of design candidates. The final proposed geometry is capable of showing better performance in terms of fulfilling the environmental requirements for the cell culture under fixed operating conditions. In this study, the explained procedures run sequentially under supervision of an interface developed in MATLAB® R2014b (MathWorks, Natick, Massachusetts, MA, United States). Figure 5 shows the scheme of the developed interface. The presented algorithm automates the creation of a list of candidate designs based on a random distribution of Δ. The most efficient design among the studied structures is identified through a 'generate and test' strategy. The illustrated algorithm delivers a simple but efficient approach to directly benefit from the capability of the CFD simulation to improve the scaffold design. The interface provides an interactive communication platform between the CFD related software

(ICEM CFD and ANSYS CFX® 15.0) (Ansys Inc., Canonsburg, Pennsylvania, PA, USA) and the result processing software (Excel 2013). Regarding the numerical simulation requirements, the studied volumes were unstructurally discretized with the help of 'Robust Octree' meshing approach, with a global element seed size equal to 10. The codes are provided as a Supplementary Materials.

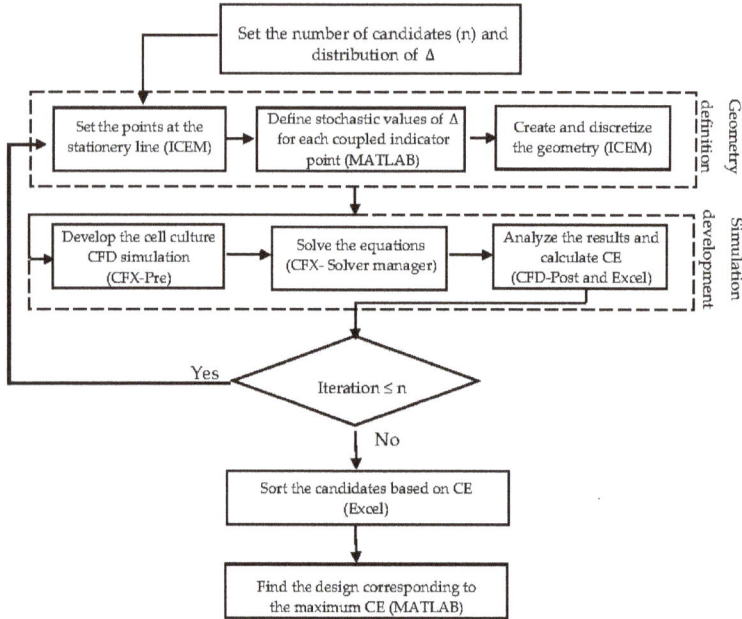

Figure 5. A schematic view of the stochastic semi-optimization routine.

4. Results and Discussion

As previously discussed, the velocity profile is a deterministic parameter in a cell culture under perfusion flow. The growing population of cells reflects the influence of the perfusion velocity during a culture. Figure 6 shows a representative simulation result of cartilage cell cultures for a range of inlet velocities and glucose concentrations within a sample scaffold unit.

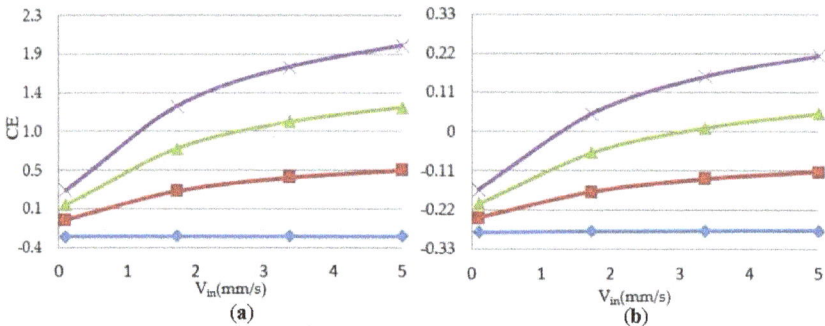

Figure 6. Correlation between cell culture efficiency vs. inlet perfusion velocity within the studied unit. Gl_{in} = ◆ 0.059×10^{-3} g/cm^3, ■ 1.54×10^{-3} g/cm^3, ▲ 3.02×10^{-3} g/cm^3, ✶ 4.5×10^{-3} g/cm^3 (a) initial cell density = 2×10^7 cells/cm^3; (b) initial cell density = 10×10^7 cells/cm^3.

The simulation results indicate the critical effect of fluid velocity on the cell culture efficiency, particularly in the case of high cell densities. Negative culture efficiency is interpreted as losing part of the cell population with time, under the specified operating conditions. The sensitivity of the cell culture to the inlet velocity profile is predicted to decrease at higher velocities. This behavior is explained by the multi-facial effect of shear stress on the cell growth rate (see Equation (8)). On the other hand, the sensitivity of the process to the flow velocity is a function of cell population. The populated cell culture shows higher resistance to the penetration of medium into the biomass volume, which overwhelms the intense influence of velocity on the cell culture.

The velocity profile in the specified unit is also a strong function of the scaffold design. In order to individually investigate the geometric effect of each pair of fibers on the fluid dynamics of the cell culture environment, two sets of simulation settings were considered. Table 2 gives an overview of the simulated configurations and the results:

The first group of simulations was initialized by the presence of a homogeneous biomass domain on the surface of the bottom fiber with an average thickness of one cartilage cell. In order to study the individual effect of each pair of fibers on the cell culture, the left-right pair (non-cultured fibers) was initially replaced by flat walls with setting Δ_3 and Δ_4 to 0. Transient simulation results of cartilage cell cultures after 10 h real time are shown in Figure 7. In this figure, the influence of the cultured fiber characteristics on the velocity profile and glucose distribution in biomass domain is shown for two different designs. The reported CFD results in this section were achieved by choosing the 'Second Order Backward Euler' solver with a constant time step (1 h), which is below the time dependency threshold of the simulation.

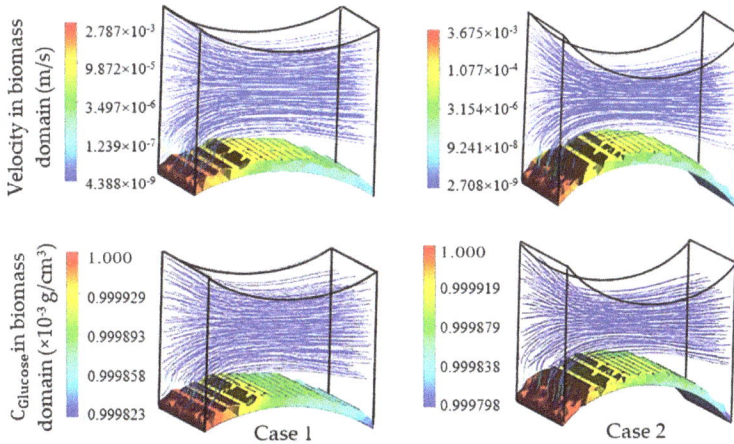

Figure 7. Physical effect of fiber 1 on the cell culture with initial cell density = 5×10^7 cells/cm^3 in the biomass phase, inlet velocity = 3 mm/s and inlet glucose concentration = 1×10^{-3} g/cm^3, $D_1^{case1} = 520$ μm, $D_2^{case1} = 0$ μm and $D_1^{case2} = 250$ μm, $D_2^{case2} = 0$ μm.

As mentioned previously, a reduction in mass transfer and nutrient supply is a direct consequence of the velocity drop at the rear side of the fibers. Although, the difference between the maximum and minimum concentrations in both designs of the unit was estimated to be small, this difference can potentially reach a significant value when the complete scaffold is considered. The curvature corresponding to the present part of the fiber in a unit is named as the effective curvature. The effective curvature is determined by a given fiber's diameter, the attachment angle and the size of the imaginary cube (Figure 5). The simulation results indicate the influence of the effective curvature on the fluid profile. As Figure 7 shows, an increase in the level of effective curvature from case 1 to case 2 results in a higher velocity drop in the downstream. An intense velocity gradient in perfused medium is directly reflected in the mass transfer rate profile, which consequently increases the possibility of ending with a significant level of heterogeneity in the biomass distribution. This result indicates a challenge in achieving a homogeneous cell distribution within a regenerated tissue. However, the heterogeneity issue can be seen as a potential in another perspective, such as for controlling cell density distribution within a scaffold by using various perfusion velocity profiles. Histology studies of natural cartilage tissue indicated the presence of distinctive zones with respect to cell distribution and morphology [33]. Usage of various scaffold designs gives the possibility of controlling the cell distribution within an engineered tissue in order to mimic the natural tissue [16].

The specified curvature also has a direct influence on the level of induced shear stress by the flow. Simulation results predict a relatively higher maximum shear stress in biomass domain for higher curvature (0.16 Pa in case 1 and 0.4 Pa in case 2), while the average parameter indicates a lower value in the same design (average shear stress in biomass phase = 0.013 Pa in case 1 and 0.011 Pa in case 2). Since for a cell culture the average value of shear stress has a higher level of importance compared to its local maximum value, a lower curvature of the cultured surface is predicted to be more desirable. On the other hand, the available surface area for the cell attachment is a critical factor in the design of an efficient cell culture structure [34,35]. The cell attachment surface area (A) in the specified unit is a direct function of the effective curvature. Hence, a larger surface area is calculated for higher curvature in case 2 (A = 4.026×10^{-4} cm^2 for case 1 and A = 4.93×10^{-4} cm^2 for case 2). Accordingly, a higher cultured curvature can provide more surface area for the cell attachment.

Similarly, the second group of simulations was initialized in the presence of a biomass domain on the surface of the bottom fiber while including the other two fibers in the simulation unit. Accordingly, two different design characteristics for the studied unit are shown in Figure 8. These simulations aim at studying the effect of the second pair of fibers as non-cultured fibers on the velocity and shear stress profile. The induced fibers mainly have an effect on the fluid dynamics of the perfusion flow by adjusting the cross-sectional area. Smaller cross-sectional areas regarding the position and characteristics of four fibers result in a higher perfusion velocity and shear stress in the unit. Transient simulation of a cartilage cell culture after 10 h real time indicates 0.08 Pa average shear stress in case 3 with 2.357×10^{-8} m^2 minimum cross-sectional area compared to 0.18 Pa in case 4, with a 1.240×10^{-8} m^2 minimum cross-sectional area. A decrease in the cross-sectional area is a beneficial factor to increase the shear stress within the positive range of effect. However, simultaneously the chance of blocking the unit increases during the cultivation.

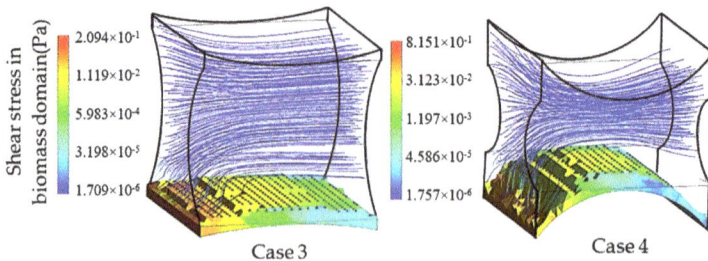

Figure 8. Hydrodynamic effect of the fiber location on the cell culture with initial cell density = 5×10^7 cells/cm^3 in the biomass phase, inlet velocity = 3 mm/s and inlet glucose concentration = 4.5×10^{-3} g/cm^3, $D_1^{case3} = 520$ µm, $D_2^{case3} = 340$ µm and $D_1^{case4} = 250$ µm, $D_2^{case4} = 145$ µm.

The efficiency of the presented method to design an improved fibrous scaffold was investigated by assuming that the biomass domain is only present on the surface of one fiber in the studied unit. This assumption gives a better understanding of the scaffold geometric effect on cell culture efficiency while the potential of the routine in a specified case study is evaluated. Using this assumption gives the possibility of dip investigation on the effect of not only cultured surfaces, but also the possible nonculture ones on controlling the cell culture environment. The presence of nonculture surfaces could potentially be considered, for instance, to mimic the multi-zonal morphology of a tissue. According to the common design of the scaffold, the initial diameter of all 4 fibers was set to 200 μm. After studding 300 design candidates, the improved geometry was obtained with displacement values of $|\Delta Y| = 7$ μm for points 1 and 2, and $|\Delta Z| = 90$ μm for points 3 and 4, which are equivalent to the improved dimensions $D_1 = 1435$ μm and $D_2 = 182$ μm for fiber 1 and 2 respectively. These values also indicate an attachment angle of $\sim 0°$ between the two fibers. A transient simulation of the cartilage cell culture for 3 mm/s inlet fluid velocity and 4.5×10^{-3} g/cm^3 glucose concentration within the improved structure is shown in Figure 9. The simulation is initialized by considering a monolayer biomass domain attached to the bottom fiber with constant cell density 5×10^7 cell/cm^3.

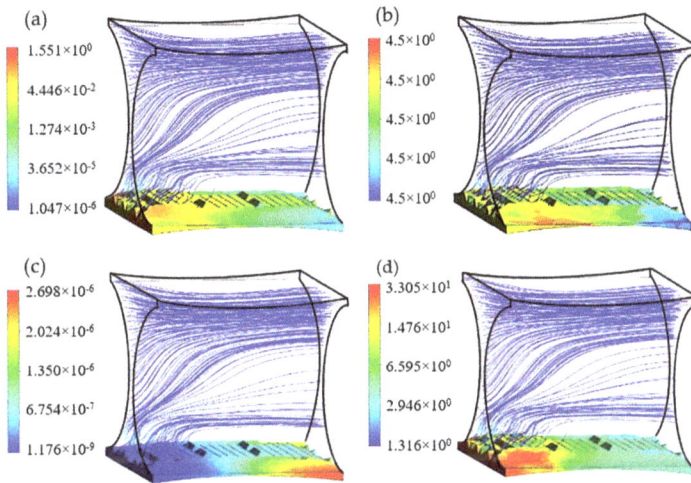

Figure 9. Cartilage cell culture within the improved geometry after 30 days cultivation with initial cell density $= 5 \times 10^7$ cells/cm^3 (**a**) Shear stress (Pa); (**b**) Glucose concentration (kg/m^3); (**c**) Lactate concentration (kg/m^3) and (**d**) Cell density $\times 10^7$ (cells/cm^3) in biomass domain.

Among the studied candidates, this geometry delivers the best performance in terms of nutrient supply balanced with metabolic growth rate, available attachment surface (A $= 4.01 \times 10^{-8}$ m^2), and flow induced average shear stress ($\bar{\tau} = 0.17$ Pa). The relatively large radius of the cultured fiber guarantees an adequate level of effective curvature for the passing flow to keep the capacity of mass transport approximately constant throughout the structure. On the other hand, the second pair of fibers has the main influence on adjusting the level of shear stress on the cell culture by optimizing the cross-sectional area for perfusion flow. In this case study, the final average cell population was predicted to reach 11.68×10^7 cells/cm^3 after 30 days cultivation, which shows up to 24% improvement compared to the reference design ($D_1 = D_2 = 200$ μm).

This method was also applied to a more realistic case study in which the attachment of the biomass phase over the entire fiber surfaces was considered. After studding 200 design candidates, the improved characteristic parameters were achieved accordingly: $D_1 = 605$ μm, $D_2 = 204$ μm and $\alpha = 20°$. The improved dimensions are expressed by displacement values $|\Delta Y| = 17$ μm for point 1 and 2, and $|\Delta Z| = 40$ μm for point 3 and 4. This scaffold can be assembled by various fabrication methods and new outstanding technologies such as advanced 3D-printing techniques. The simulation results of 200 design candidates are shown in Figure 10a. The collective results indicate an inverse correlation between the fiber's diameters ratio D_1/D_2 (proportional to the corresponding value $\Delta Y/\Delta Z$) and the cell culture efficiency.

The simulation results in Figure 10b show a considerable risk of negative culture efficiency for $|\Delta Y / \Delta Z| > 5.5$. Moreover, designs in the range of $|\Delta Y / \Delta Z| \in [0–1]$ are expected to deliver the highest cell culture efficiencies within the studied range. This range is classified into two major regions E_1 and E_2 in Figure 10b. Although the design criterion $\left| \frac{\Delta Y}{\Delta Z} \right| \in E_1$ was predicted to deliver a more appropriate culture environment, the design criterion $\left| \frac{\Delta Y}{\Delta Z} \right| \in E_2$ proposes more feasible designs when considering the ease of fabrication.

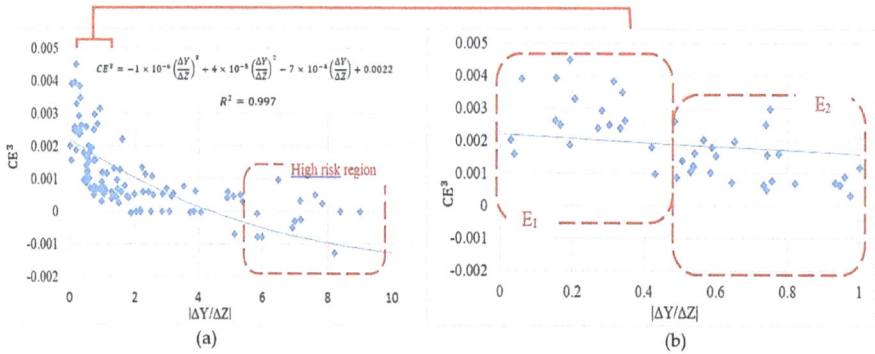

Figure 10. The effect of scaffold design on the cartilage cell culture efficiency, (**a**) full simulation results; (**b**) simulation results in a range of $|\Delta Y / \Delta Z| \in [0–1]$.

Table 2. Overview of the simulation exploratory workspace for an improved scaffold design.

Figure	Case	Case Configuration	Diameters	Result
3	Reference	Biomass present on surface of all four fibers	$D_1 = D_2 = 200 \ \mu m$	
7	Case 1 Case 2	Biomass present only on surface of one fiber from the pair 1. Presence of 1 pair of fibers in the unit	The cultured fiber: $D_1^{case1} = 520 \ \mu m$ $D_1^{case2} = 250 \ \mu m$ Non_cultured pair $D_2 = 0 \ \mu m$	$EC_{case2} > EC_{case1}$ * A higher velocity drop at the rear side of the cultured fiber in case 2 A higher maximum shear stress and lower average shear stress in case 2 A higher cell attachment area in case 2
8	Case 3 Case 4	Biomass present only on surface of one fiber from the pair 1. Presence of 2 pairs of fibers in the unit	The cultured fiber: $D_1^{case3} = 520 \ \mu m$ $D_1^{case4} = 250 \ \mu m$ $D_1 \neq 0 \ \mu m$ Non_cultured pair: $D_2^{Case\ 3} > D_2^{Case\ 4}$ $D_2 \neq 0 \ \mu m$	$EC_{case4} > EC_{case3}$ * A higher velocity drop at the rear side of the cultured fiber in case 4 A higher cell attachment area in case 2 $A_3 > A_4$ ** A lower flow cross-sectional area in case 4 A higher level of shear stress in case 4
9	Improved design	Biomass present only on surface of one fiber from the pair 1. Presence of 2 pairs of fibers in the unit	The cultured fiber $D_1 = 1435 \ \mu m$ Non_cultured pair: $D_2 = 185 \ \mu m$	24% improvement is predicted compared with the reference design

* EC: Effective curvature of the cultured surface; ** A: minimum intersection area for the perfusion flow.

5. Conclusions

Nutrient supply, metabolite concentrations and flow-induced hydrodynamic conditions are the main controlling parameters for cartilage cell culture in a perfusion bioreactor [3,13,20]. Using mathematical modelling and CFD simulation helps to predict the state of the cell culture during the process with respect to effective parameters such as perfusion flow rate, geometric design, influent glucose concentration etc.

In the present study, the potential of modelling and CFD simulation to design more efficient cell culture processes was investigated. Hence, a complex multiphase model, including the fluid flow and the growing cells, was presented to predict the state of the culture during a cartilage cell culture under a perfusion flow. In this model, the effects of glucose concentration, shear stress and pH of the environment on the cells were included. It should be considered that this model does not cover the effect of other involved nutrients such as oxygen on the cell metabolism, which requires further investigation. However, the developed model and the corresponding CFD simulation provide valuable information about the requirements, challenges and considerations for an efficient cell culture from a scaffold design point of view. In order to take advantage of this capability, a stochastic algorithm was introduced to evaluate various structure candidates, and to select the most efficient one. Based on that, an automated routine was developed in order to study various geometries virtually, within a simulation software environment, and to assess new, potentially useful structures for the cell culture. Here, a frequently used fibrous scaffold was implemented as a case-study. The proposed improved design has demonstrated to deliver significantly higher efficiency compared with the reference design in terms of mass transfer rate and hydrodynamic parameters. As the result, up to 24% improvement in the final cell population was predicted for the new design.

This promising approach can be considered for other types of scaffolds by numerically evaluating different geometrical structures within a relatively shorter operation time, while the experimental effort is mostly limited to the validation of the model.

Supplementary Materials: The following are available online at http://www.mdpi.com/2306-5354/5/2/33/s1.

Author Contributions: T.T. produced the results, collected the information and wrote the manuscript. M.J.A. has partly conceived the research. K.V.G. and U.K. have conceived the research, provided feedbacks to the content and participated in writing the manuscript.

Conflicts of Interest: The authors declare no conflict of interest.

Nomenclature

A	surface area [m^2]
$c_g{}^f$	average glucose concentration in fluid [kg/m^3]
$c_g{}^c$	average glucose concentration in cell [kg/m^3]
C_l	lactate concentration [mol/m^3]
D	fiber's diameter [m]
d_c	thickness of the biomass domain [mm]
$D_g{}^f$	diffusion coefficient of glucose in fluid [cm^2/s]
$D_g{}^c$	diffusion coefficient of glucose in cell [cm^2/s]
K_c	saturation coefficient
K_{g0}	maximum growth rate under static culture conditions [1/s]
r_g	cell growth rate [1/s]
r_d	cell death rate [1/s]
R_g	Glucose consumption rate [kg/m^3 s]
V_m	maximum glucose consumption rate [kg/m^3 s]
K_m	half-saturation constant for glucose [kg/m^3]
ε_f	fluid volume fraction
ε_c	cell volume fraction
$\rho_c{}^b$	average biomass density [g/cm^3]

ρ_{cell}	cellular density [g/cm^3]
τ	shear stress [Pa]
Δ	moving point displacement value [μm]
α	attachment angle [Rad]
h_c	compaction depth [μm]
μ_c	local viscosity of the biomass domain [Pa. s]

References

1. Buckwalter, J.A.M. Articular Cartilage Injuries. *Clin. Orthop. Relat. Res.* **2002**, *402*, 21–37. [CrossRef]
2. Flugge, L.A.; Miller-Deist, L.A.; Petillo, P.A. Towards a molecular understanding of arthritis. *Chem. Biol.* **1999**, *6*, R157–R166. [CrossRef]
3. Freed, L.E.; Marquis, J.C.; Langer, R.; Vunjak-Novakovic, G. Kinetics of chondrocyte growth in cell-polymer implants. *Biotechnol. Bioeng.* **1994**, *43*, 597–604. [CrossRef] [PubMed]
4. Pazzano, D.; Mercier, K.A.; Moran, J.M.; Fong, S.S.; Di Biasio, D.D.; Rulfs, J.X.; Kohles, S.S.; Bonassar, L.J. Comparison of Chondrogensis in Static and Perfused Bioreactor Culture. *Biotechnol. Prog.* **2000**, *16*, 893–896. [CrossRef] [PubMed]
5. Chaudhuri, J.; Al-Rubeai, M. *Bioreactors for Tissue Engineering*; Springer: Berlin, Germany, 2005.
6. Wang, J.H.-C.; Thampatty, B.P. An Introductory Review of Cell Mechanobiology. *Biomech. Model. Mechanobiol.* **2006**, *5*, 1–16. [CrossRef] [PubMed]
7. Hutmacher, D.W.; Singh, H. Computational fluid dynamics for improved bioreactor design and 3D culture. *Trends Biotechnol.* **2008**, *26*, 166–172. [CrossRef] [PubMed]
8. International Society of Biorheology; Trindade, M.C.D.; Ikenoue, T.; Mohtai, M.; Das, P.; Carter, D.R.; Goodman, S.B.; Schurman, D.J. *Biorheology*; Pergamon Press: Oxford, UK, 2000; Volume 37.
9. Williams, K.A.; Saini, S.; Wick, T.M. Computational Fluid Dynamics Modeling of Steady-State Momentum and Mass Transport in a Bioreactor for Cartilage Tissue Engineering. *Biotechnol. Prog.* **2002**, *18*, 951–963. [CrossRef] [PubMed]
10. Raimondi, M.T.; Boschetti, F.; Falcone, L.; Migliavacca, F.; Remuzzi, A.; Dubini, G. The effect of media perfusion on three-dimensional cultures of human chondrocytes: Integration of experimental and computational approaches. *Biorheology* **2003**, *41*, 401–410.
11. Krühne, U.; Wendt, D.; Martin, I.; Juhl, M.V.; Clyens, S.; Theilgaard, N. A Transient 3D-CFD Model Incorporating Biological Processes for Use in Tissue Engineering. *Micro Nanosyst.* **2010**, *2*, 249–260.
12. Nava, M.M.; Raimondi, M.T.; Pietrabissa, R. A multiphysics 3D model of tissue growth under interstitial perfusion in a tissue-engineering bioreactor. *Biomech. Model. Mechanobiol.* **2013**, *12*, 1169–1179. [CrossRef] [PubMed]
13. Sacco, R.; Causin, P.; Zunino, P.; Raimondi, M.T. A multiphysics/multiscale 2D numerical simulation of scaffold-based cartilage regeneration under interstitial perfusion in a bioreactor. *Biomech. Model. Mechanobiol.* **2011**, *10*, 577–589. [CrossRef] [PubMed]
14. Hollister, S.J. Porous scaffold design for tissue engineering. *Nat. Mater.* **2005**, *4*, 518–524.
15. Chen, Y.; Zhou, S.; Li, Q. Microstructure design of biodegradable scaffold and its effect on tissue regeneration. *Biomaterials* **2011**, *32*, 5003–5014. [CrossRef] [PubMed]
16. Woodfield, T.B.F.; Blitterswijk, C.A.; Van Wijn, J.; De Sims, T.J.; Hollander, A.P.; Riesle, J. Polymer Scaffolds Fabricated with Pore-Size Gradients as a Model for Studying the Zonal Organization within Tissue-Engineered Cartilage Constructs. *Tissue Eng.* **2005**, *11*, 1297–1311. [CrossRef] [PubMed]
17. Hossain, M.S.; Bergstrom, D.J.; Chen, X.B. Modelling and simulation of the chondrocyte cell growth, glucose consumption and lactate production within a porous tissue scaffold inside a perfusion bioreactor. *Biotechnol. Rep.* **2015**, *5*, 55–62. [CrossRef] [PubMed]
18. Yu, J.; Hu, K.; Smuga-Otto, K.; Tian, S.; Stewart, R.; Slukvin, I.I.; Thomson, J.A. Human induced pluripotent stem cells free of vector and transgene sequences. *Science* **2009**, *324*, 797–801. [CrossRef] [PubMed]
19. Bilgen, B.; Barabino, G.A. Location of scaffolds in bioreactors modulates the hydrodynamic environment experienced by engineered tissues. *Biotechnol. Bioeng.* **2017**, *98*, 282–294. [CrossRef] [PubMed]
20. Bibby, S.R.S.; Fairbank, J.C.T.; Urban, M.R.; Urban, J.P.G. Cell Viability in Scoliotic Discs in Relation to Disc Deformity and Nutrient Levels. *Spine* **2002**, *27*, 2220–2227. [CrossRef] [PubMed]

21. Horner, H.A.M.P.; Urban, J.P.G. 2001 Volvo award winner in basic science studies: effect of nutrient supply on the viability of cells from the nucleus pulposus of the intervertebral disc. *Spine* **2001**, *26*, 2543–2549. [CrossRef] [PubMed]

22. Wood, B.D.; Quintard, M.; Whitaker, S. Calculation of effective diffusivities for biofilms and tissues. *Biotechnol. Bioeng.* **2002**, *77*, 495–516. [CrossRef] [PubMed]

23. Lin, T.-H.; Jhang, H.-Y.; Chu, F.-C.; Chung, C.A. Computational modeling of nutrient utilization in engineered cartilage. *Biotechnol. Prog.* **2013**, *29*, 452–462. [CrossRef] [PubMed]

24. Zhou, S.; Cui, Z.; Urban, J.P.G. Nutrient gradients in engineered cartilage: Metabolic kinetics measurement and mass transfer modeling. *Biotechnol. Bioeng.* **2008**, *101*, 408–421. [CrossRef] [PubMed]

25. Heywood, H.K.; Heywood, H.K.; Lee, D.A.; Oomens, C.W.; Bader, D.L. Nutrient Utilization by Bovine Articular Chondrocytes: A Combined Experimental and Theoretical Approach. *J. Biomech. Eng.* **2005**, *127*, 758. [CrossRef]

26. Chung, C.A.; Yang, C.W.; Chen, C.W. Analysis of cell growth and diffusion in a scaffold for cartilage tissue engineering. *Biotechnol. Bioeng.* **2006**, *94*, 1138–1146. [CrossRef] [PubMed]

27. Galban, C.J.; Locke, B.R. Analysis of cell growth kinetics and substrate diffusion in a polymer scaffold. *Biotechnol. Bioeng.* **1999**, *65*, 121–132. [CrossRef]

28. Chung, C.A.; Chen, C.W.; Chen, C.P.; Tseng, C.S. Enhancement of cell growth in tissue-engineering constructs under direct perfusion: Modeling and simulation. *Biotechnol. Bioeng.* **2007**, *97*, 1603–1616. [CrossRef] [PubMed]

29. Shakhawath Hossain, M.; Bergstrom, D.J.; Chen, X.B. A mathematical model and computational framework for three-dimensional chondrocyte cell growth in a porous tissue scaffold placed inside a bi-directional flow perfusion bioreactor. *Biotechnol. Bioeng.* **2015**, *112*, 2601–2610. [CrossRef] [PubMed]

30. Tajsoleiman, T.; Abdekhodaie, M.J.; Gernaey, K.V.; Krühne, U. Geometry optimization of a fibrous scaffold based on mathematical modelling and CFD simulation of a dynamic cell culture. *Comput. Aided Chem. Eng.* **2016**, *38B*, 1413–1418.

31. Galban, C.J.; Locke, B.R. Effects of spatial variation of cells and nutrient and product concentrations coupled with product inhibition on cell growth in a polymer scaffold. *Biotechnol. Bioeng.* **1999**, *64*, 633–643. [CrossRef]

32. Chung, C.A.; Ho, S.-Y. Analysis of Collagen and Glucose Modulated Cell Growth within Tissue Engineered Scaffolds. *Ann. Biomed. Eng.* **2010**, *38*, 1655–1663. [CrossRef] [PubMed]

33. An, Y.H.; Martin, K.L. *Handbook of Histology Methods for Bone and Cartilage*; Humana Press: Totowa, NJ, USA, 2003.

34. Srivastava, V.M.L.; Malemud, C.J.; Sokoloff, L. Chondroid Expression by Lapine Articular Chondrocytes in Spinner Culture Following Monolayer Growth. *Connect. Tissue Res.* **1974**, *2*, 127–136. [CrossRef] [PubMed]

35. Grundmann, K.; Zimmermann, B.; Barrach, H.-J.; Merker, H.-J. Behaviour of epiphyseal mouse chondrocyte populations in monolayer culture. *Virchows Arch. A Pathol. Anat. Histol.* **1980**, *389*, 167–187. [CrossRef] [PubMed]

MDPI

St. Alban-Anlage 66

4052 Basel

Switzerland

Tel. +41 61 683 77 34

Fax +41 61 302 89 18

www.mdpi.com

Bioengineering Editorial Office

E-mail: bioengineering@mdpi.com

www.mdpi.com/journal/bioengineering

www.ingramcontent.com/pod-product-compliance
Lightning Source LLC
Chambersburg PA
CBHW041216220326
41597CB00033BA/5989